EXAMINING THE EXAMINATIONS

An International Comparison
of Science and Mathematics
Examinations for
College-Bound Students

Evaluation in Education and Human Services

Editors:

George F. Madaus, Boston College,
  Chestnut Hill, Massachusetts, U.S.A.
Daniel L. Stufflebeam, Western Michigan
  University, Kalamazoo, Michigan, U.S.A.

Other books in the series:

EXAMINING THE EXAMINATIONS

An International Comparison of Science and Mathematics Examinations for College-Bound Students

edited by

Edward D. Britton
Senta A. Raizen

from

**The National Center for Improving
Science Education**

a division of
The NETWORK, Inc.

KLUWER ACADEMIC PUBLISHERS
Boston/Dordrecht/London

Distributors for North America:
Kluwer Academic Publishers
101 Philip Drive
Assinippi Park
Norwell, Massachusetts 02061 USA

Distributors for all other countries:
Kluwer Academic Publishers Group
Distribution Centre
Post Office Box 322
3300 AH Dordrecht, THE NETHERLANDS

Library of Congress Cataloging-in-Publication Data
ISBN-13:978-94-010-6648-8   e-ISBN-13:978-94-009-0363-0
DOI: 10.1007/978-94-009-0363-0

A C.I.P. Catalogue record for this book is available from
the Library of Congress.

The National Center for Improving Science Education

The National Center for Improving Science Education (NCISE) is a division of The NETWORK, Inc., a nonprofit organization dedicated to educational reform. The Center's mission is to promote change in state and local policies and practices in science curriculum, teaching, and assessment. We carry out our mission by providing a range of products and services to educational policymakers and practitioners who work to strengthen science teaching and learning all across the country.

Our products and services are oriented toward practical guidance for those responsible for the day-to-day decisions that shape diverse learning environments, preschool to postsecondary. We are dedicated to helping all stakeholders in science education reform, including the postsecondary institutions that train prospective science teachers, to promote better science education for all students.

NCISE Advisory Board Members

Contents

Part One:
Looking Inside the Examinations

Part Two:
Understanding the Examination Systems

List of Figures

List of Tables

Contributors

The Editors

Edward D. Britton is an associate director of the National Center for Improving Science Education (NCISE) and serves as project director for several international studies, recently contributing to *What College-Bound Students Are Expected to Know About Biology*. He helped design the curriculum analysis and science framework of the Third International Mathematics and Science Study (TIMSS). As part of multinational research led by the Organization for Economic Co-operation and Development (OECD), Britton is coordinating case studies of eight major US innovations in science and mathematics education. He has written on several aspects of K-16 science education, including indicators for science education, dissemination of innovations, curriculum studies, and evaluation. Britton also has managed development of CD-ROM disks and videotapes designed to help elementary teachers enhance their science knowledge and pedagogy.

Senta A. Raizen is the director of NCISE. She is the primary author of several NCISE reports on science education in elementary, middle, and high school and books on indicators in science education, preservice education of elementary school teachers, and technology education. In addition to science education, Raizen's work includes educational assessment and program evaluation, education policy, reforming education for work, and linking education research and policy with practice. She is principal investigator for NCISE's evaluations of several federal programs that support science education, the TIMSS study, and the OECD case study project, for which she is working on a synthesis of 23 case studies of innovations in science, mathematics, and technology education produced by over a dozen industrialized countries. She is a member of the International Steering Committee for TIMSS and serves in an advisory capacity to several national education studies, including the National Assessment of Educational Progress (NAEP), the National Goals Panel, and the National Institute for Science Education.

The Authors

John Dossey (Chapter 6), Distinguished University Professor of Mathematics at Illinois State University, is a past president of the National Council of Teachers of Mathematics. He has participated in the US aspects of the Second International Mathematics Study and the NAEP. At present, Dossey is chair of the US National Commission of Mathematical Instruction and chair of the Conference Board of the Mathematical Sciences.

Dwaine and Lucy Eubanks (Chapter 4), both at Clemson University, work collaboratively. Dwaine Eubanks directs the Examinations Institute of the American Chemical Society's (ACS's) Division of Chemical Education. Lucy Eubanks is the Institute's associate director. Each has chaired the ACS Division of Chemical Education, and both have written college-level chemistry textbooks. They co-direct a computer-based assessment project funded by the National Science Foundation (NSF).

Matthew Gandal (Chapters 7 and 8), senior associate in the Educational Issues Department of the American Federation of Teachers, is principal author of *What College-Bound Students Abroad Are Expected to Know About Biology and Making Standards Matter, A Fifty-State Progress Report on Efforts to Raise Academic Standards*. He has written and spoken in other forums about academic standards and school reform.

Kjell Gisselberg (Chapter 5), University of Umeå, chairs the Swedish National Committee on National Tests in Physics. He is the National Research Coordinator for TIMSS in Sweden. He contributed a chapter on the Swedish national tests in physics to a UNESCO publication, *Physics Examinations for University Entrance*.

Simon Hawkins (Chapters 2, 7, and 8), research associate at NCISE, is a contributing author to *What College-Bound Students Abroad Are Expected to Know About Biology* and a principal author of the NCISE report, *How Elementary School Teachers Are Prepared in Science: Reporting on 142 Preservice Programs*. He has also assisted in research projects on professional development and technology education.

Pinchas Tamir (Chapter 3), holds the Charles and Marylin Gold Family Chair in Education at Hebrew University of Jerusalem and is the TIMSS coordinator for Israel. For the past 20 years, with funding from NSF, he has worked on developing innovative student assessments in science and evaluated the feasibility of using such assessments in the classroom.

The Reviewers

Paul Black (all chapters, particularly Chapter 5), King's College London, University of London, was editor of a UNESCO study of physics examinations in 11 countries, *Physics Examinations for University Entrance*. He is past president of the UK Association for Science Education and current chair of the International Commission on Physics Education.

Angelo Collins (Chapter 3), Vanderbilt University, was director of the National Science Education Standards Project coordinated by the National Academy of Sciences/National Research Council and director of the Teacher Assessment Project at Stanford University. Her research interests include understanding and teaching genetics, alternative assessments for science students and teachers, and science teacher education.

Dorothy Gabel (Chapter 4), Indiana University, is a past president of the National Association for Research in Science Teaching, the School and Mathematics Association, and the Hoosier Association of Science Teachers. She is editor of the *Handbook of Research in Science Teaching and Learning*, author of a high school chemistry textbook and problem-solving book, and director of a chemistry videotape project on effective chemistry teaching.

Curtis McKnight (Chapter 6), Mathematics Department, University of Oklahoma, is a principal researcher in the TIMSS curriculum analysis. He is lead author of *The Underachieving Curriculum*, the US report on the Second International Mathematics Study. In addition to large-scale, cross-national studies, his research includes cognitive and information-processing studies.

David Robitaille (general), University of British Columbia, is head of the Department of Curriculum Studies and professor of Mathematics Education. His research has focused on large-scale assessment of students' achievement in mathematics at the provincial, national, and international levels. Robitaille is the international coordinator of TIMSS.

John Schwille (Chapters 7 and 8), Michigan State University, is professor and assistant dean for international studies in education. He specializes in comparative education and the sociology of education, with an emphasis on comparative understanding of teaching and learning as influenced by their institutional and organizational contexts.

FOR YEARS, REFORM RHETORIC has proclaimed that US education should be second to none. But what exactly does that mean, given the lack of detailed information available on other nations' curricula? While previous comparative studies have investigated critical features of educational systems around the world—including their examination systems and courses of study—these analyses have not captured the specifics. What topics do teachers actually teach from day to day? What instructional practices do they follow? What are the characteristics of textbooks and other instructional materials? What are important examinations like? Some research has addressed curricular materials and instructional practices. These studies generally are limited to one or two countries: systematic, multinational comparisons of these detailed aspects of education are rare.

Large-scale international assessments have been conducted, however, to check relative progress in student achievement; these include the First and Second International Mathematics Studies, the First and Second International Science Studies, the ongoing Third International Mathematics and Science Study (TIMSS), and the International Assessment of Educational Progress. Past research thus has measured outcomes repeatedly and extensively, but has not given a similar level of attention to understanding the curricular contrasts from which outcome differences arise. One significant exception is TIMSS; This 1993-97 study includes a detailed, empirical curriculum analysis along with a thorough assessment of student learning. (See "Connections to TIMSS," in Chapter 1.)

Examining the Examinations takes a fine-grained look at the required advanced science (biology, chemistry, and physics) and mathematics examinations taken by college-bound students in seven countries. The scholarly appraisal presented here focuses on topics covered, types of questions used, and performances expected from students. To generate the results included in this book, scientists and mathematicians pored over every question of every examination studied and also considered each examination as a whole. For example, the physics chapter alone reports the analysis of 1,300 examination questions. This book also gives new detail on examination systems, and discusses the many frames of reference that could and should be considered when comparing the "difficulty" of examinations.

The book concentrates on comparisons of the examinations, illustrating their similarities and differences with selected questions taken from the actual examinations. Readers seeking more examples of examination questions can consult several sources. Notably, in a companion project to the present effort, the National Center for Improving Science Education (NCISE) and the American Federation of Teachers (AFT) copublished *What College-Bound Students Are Expected to Know About Biology: Exams from England and Wales, France, Germany and Japan* (Gandal et al. 1994).[1] This publication contains recent (1991-92), complete examinations from several countries, including the United States. Additionally, Black (1992) incorporates a few sample questions of physics examinations from 11 countries; Wu (1993) presents 10 Japanese mathematics examinations; and Eckstein and Noah (1993) include sample sections on mathematics, language, and literature from several countries' examinations administered between 1988 and 1990. None of these, however, provides a scholarly comparison of examinations across subjects and countries; that is the contribution of *Examining the Examinations.*

This analysis provides insight into international examinations, and hence into the curricula associated with them. The very nature of these challenging, high-stakes examinations means that they drive—to some degree—the educational systems of their respective countries. Thus, the examinations heavily influence what topics are taught and, to a lesser extent, how they are taught.

Should the United States consider having examinations like those of other countries? NCISE is not taking a position on this matter since many scholars—as well as the mass media—already are engaged in that debate. But we do urge strongly that more substance be put into that discussion. Beneath the surface of whether the United States should or even could have examinations comparable to those abroad are a host of particulars that should be considered.

The international comparisons presented in this book offer us a window on educational "laboratories" in six other countries. The examinations in other countries can be viewed as experiments based on different methods than those typically used in this country. While there are hazards and limitations inherent in transferring one nation's strategies to another country's context, issues of transferability should not discourage researchers and policymakers from looking outside US borders for fresh ideas and perspectives. We hope this book is useful, not only for considering questions that readers already have in mind, but for prompting new ones.

---

[1] Available at cost from AFT. Also, a 24-page excerpt was published in the *American Educator* (AFT, 1994). NCISE and AFT will publish two similar volumes in late 1995 and early 1996, one on chemistry and physics examinations; the other on mathematics examinations.

Audience

Examining the Examinations has been designed with several audiences in mind. It offers policymakers and educational organizations at the national, state, and local levels a rare substantive source to inform the national debate on what constitutes curricula of international caliber, particularly the controversy over whether to have national end-of-secondary school examinations in the United States and, if so, what their features might be. Science and education organizations will want to judge the fit between their views of what topics constitute scientific literacy and the examinations' topics. Scientists may consider what these examinations imply about the entry-level knowledge of college students in other nations. Teacher educators will want to consider the preparation that other nations' teachers must receive for their roles in equipping students to take examinations and in constructing and grading these primarily free-response examinations. Faculty concerned with teacher preparation programs will be interested in the depth of science or mathematics knowledge expected of secondary teachers in other countries. And education researchers and educators will want to consider all of the above issues and more—for example, what kinds of learning experiences would be needed to prepare students for such examinations.

Organization

The two parts of *Examining the Examinations* parallel the main emphases of the study. Part I discusses the principal research focus: internal characteristics of examinations. The secondary purpose of the study, understanding and comparing the corresponding examination systems, is treated in Part II. To help the reader understand the results presented in these two parts, Chapter 1 provides background information about the study. It briefly explains examination selection, the study's connections to TIMMS, and the approaches used to compare examinations and examination systems. Chapter 1 concludes with short descriptions of the seven examination systems.

Part I: Looking Inside the Examinations

Part I begins with Paul Black's commentary on our findings about the examinations' characteristics. Chapter 2 reports the similarities and contrasts among examinations of different subjects, from different nations, and within different examination systems. It also discusses the curricular implications of our results, including the different "lenses" that should be used when considering the difficulty of examinations. Chapters 3-6 make up the core of this study. Each describes the features of examinations in one of the three science fields—biology, chemistry, and physics—or in mathematics. Each chapter provides an overview of the examinations and compares their general structures, item characteristics, topics, and performance expectations.

Part II: Understanding the Examination Systems

Part II begins with John Schwille's commentary on the value of comparing examinations and the importance of understanding examination systems as part of such studies. Chapter 7 compares the examination systems of the countries studied in this report, noting significant similarities and differences among them. It focuses on the issues of examination purposes, number of examinees and pass rates, governance, differentiation between academic and nonacademic students, programs of study leading to examinations, creation of examinations, scoring and grading, and system and student costs. Like Chapter 2, it also describes several lenses that need to be considered when comparing the difficulty of examinations. Chapter 8 provides in-depth descriptions of each examination system, organized around the issues used for comparison in Chapter 7.

Edward D. Britton   *Washington, DC*
Senta A. Raizen   *September, 1995*

Acknowledgments

We gratefully acknowledge the many individuals who contributed to this project, from gathering the examinations to writing the final product. The authors deserve special praise for the mammoth undertaking of reviewing and analyzing every question in every examination. All the reviewers' insights strengthened the book, and Paul Black and John Schwille also wrote commentaries for Parts I and II. Barbara Buckley provided additional analyses for the biology chapters and checked the translations of the biology examinations. In addition to writing their chapters on biology and physics examinations, Pinchas Tamir and Kjell Gisselberg explained the examination systems of Israel and Sweden, respectively, for us.

David Robitaille wrote an explanation of how this book complements the Third International Mathematics and Science Study and arranged for many examination translations, as did Anne Cramer of Language Learning Enterprises. Through a productive collaboration with Matthew Gandal and others at the American Federation of Teachers, we were able to publish in full some of the examinations analyzed in this book. Nita Congress provided invaluable, thoughtful editorial assistance; Shelley Wetzel of Marketing Options handled the graphics for us. At The NETWORK, Inc., Sue Martin helped guide us through the labyrinth of book publishing.

Numerous individuals helped us obtain the examinations described in this volume, as noted in the Chapter 1 section on examination selection. The challenging endeavor of locating information about examination systems for Part II was only possible because many individuals in the United States and abroad gave generously of their time. Their efforts are acknowledged in Chapter 8. Finally, we are grateful to the National Science Foundation (NSF) for funding this work under Grant RED-9253080, and to Larry Suter at NSF for providing supportive monitoring throughout this project.

Edward D. Britton
Senta A. Raizen

Study Background

Edward Britton

Study Overview

THIS BOOK PRESENTS THE results, implications, and issues arising from a scholarly appraisal of end-of-secondary school examinations in mathematics and science (biology, chemistry, physics) for college-bound students in England and Wales, France, Germany, Israel, Japan, Sweden, and the United States. An international team of scientists and mathematicians considered every question of every examination and also reviewed each examination as a whole. They analyzed the examinations' topics, item types, and expectations for student performance, as well as general features of the examinations such as their lengths, components, and choices given to students in selecting questions. The book also includes descriptions of and contrasts among the various countries' examination systems;[1] this information provides a useful context for interpreting the examination comparisons. System characteristics discussed are the examination purposes, how they are created and scored, centralization of examination authority, the courses of study that lead to the examinations, and cost issues.

Selection of Examinations

The study reported in this book focuses on a very specific and influential determinant of other nations' curricula—examinations. In many countries, students at the end of secondary school who plan further study must take a number of examinations that probe their command of advanced topics in

---

[1] We use the term examination system to refer to the aspects of a country's educational system that pertain to examinations. What we call examination systems typically are part of or closely linked to the countries' formal educational systems. However, all aspects of the Advanced Placement examinations in the US are determined by a private company.

specific subjects. Demonstrating the high stakes attached to these examinations is the fact that in several countries the examinations used in previous years and illustrative answers are published, and students and teachers study them assiduously. The demand for old examinations is so high that in England and Wales, France, Germany, Japan, and the United States, collections of these examinations are published and sold as books. The market for past French examinations even permits multiple publishers to sell competing books successfully.

Classroom and private instruction are designed to equip students with the skills to pass these high-stakes examinations. Since these examinations greatly influence what topics are taught—and, to some extent, how they are taught—their analysis sheds light on a critical determinant of other countries' curricula.

Although many nations administer school-leaving examinations for their college-bound students, resource limitations confined our study to the United States and six other countries. England and Wales, France, Germany, and Japan were selected for inclusion in the study because they are among the major industrial countries with which the United States often compares itself. Further, the examinations from England and Wales and from France have influenced examinations in many of their former colonies. We also decided to include examinations from Sweden and Israel because they have some interesting characteristics that are quite different from those of the other nations.

Two sets of examinations were sampled from within England and Wales, and France, and Germany, because these countries do not use a single, national examination for all students. Instead, each German state has different examinations, as do various regions of France; seven private examination boards create examinations in England and Wales.

Once we had determined what countries to include in the study, we decided which of these countries' many examinations should be analyzed. Since the goal was to compare examinations taken by those secondary students most likely to study science or mathematics in college, the following criteria generally were used to select examinations:

- given at the end of secondary school
- given to college-bound students
- specific to biology, chemistry, physics, or mathematics
- perceived to be the most challenging examinations

After these criteria had been established, selecting examinations was—for most countries—fairly straightforward. There were, however, a few notable exceptions.

- Sweden's Central Test examinations, which met all the other criteria, are administered to all students, not just the college-bound.

- Japan has a government-sponsored national examination that meets the first three criteria. Nevertheless, we opted to analyze individual university entrance examinations, because they are required by more universities and are generally accepted as more advanced than the national examinations. Among the hundreds of possible university examinations, we chose those from Tokyo University, commonly held to be the most prestigious.

- In the United States, the Scholastic Achievement Test (SAT I) is the examination most widely taken by college-bound students. It does not, however, include science or assess the most difficult mathematics topics. Thus, we elected to analyze only the Advanced Placement (AP) examinations from the College Board's Educational Testing Service because they are the most challenging subject-specific examinations in the United States—even though they are not taken by many college-bound students.[2]

- France places chemistry and physics in one examination as two separate sections. The appropriate subject experts analyzed the individual sections.

Table 1-1 lists the examinations we analyzed. Because there is no single "national" examination in England and Wales, France, or Germany, examinations from two regions in France, two states (Länder) in Germany, and two examining boards in England and Wales were used. Two years of examinations were analyzed—where available—so as to gain a fuller understanding of each type of examination. For each subject (biology, chemistry, physics, and mathematics), we looked at from 17 to 22 complete examinations.[3]

---

[2] American College Testing examinations, which assess science reasoning, and the Achievement Examinations of The College Board, which assess science content knowledge, were considered similar to—but less challenging than—the AP examinations.

[3] Numerous individuals helped us obtain the examinations described in this volume. They include Derex Foxman of the National Foundation for Educational Research in England and Wales; Gordon Stobart, director of research for the London Examinations and Assessment Council; Anne Servant of the Ministere de l'Education Nationale et de la Culture in France; Rainer Lehman at the University of Hamburg in Germany; and Walter McDonald of the Educational Testing Service in the United States.

Chapters 7 and 8 describe and compare the countries' examination systems and thus shed more light on how examinations were selected for this study.

Table 1-1.   Biology, Chemistry, Physics, and Mathematics Examinations Selected for Analysis

Country	Region/Board	Examination Name	Years	Subjects
E&W	Associated Examining Board; University of London Examination and Assessment Council	Advanced Level (called "A-Level")	1991, 1992	All four subjects
France	Paris; Aix	Baccalauréat	1991, 1992	All four subjects
Germany	Bavaria; Baden-Württemberg	Abitur	1991, 1992	All four subjects
Israel	—	Bagrut	1992	No mathematics; additional biology for 1991
Japan	—	Tokyo University Entrance	1991, 1992	All four subjects; no biology 1992
Sweden	—	Central Test	1991, 1992	No biology
US	—	Advanced Placement	Varied, 1993	All four subjects 1993; mathematics and physics, 1988; biology, 1990; chemistry, 1989

Connections to TIMSS[4]

The examinations study discussed in this book is a companion project to the Third International Mathematics and Science Study (TIMSS). Over 50 countries are participating in one or more aspects of TIMSS, making it the largest

---

[4] This section was contributed by David Robitaille, international coordinator of the Third International Mathematics and Science Study.

and most ambitious international comparison of educational systems and achievement ever conducted. TIMSS is a project of the International Association for the Evaluation of Educational Achievement (IEA), which has had many years of experience in conducting international comparisons in various aspects of education. Most of the researchers in this examinations study also are involved in TIMSS, in particular with the TIMSS Population 3 level. Population 3 consists of all students enrolled in the last year of the secondary school system, including those students identified nationally as "specialists" in either mathematics or physics.

The TIMSS Curriculum Frameworks for Mathematics and Science (Robitaille et al., 1993) were developed to make it possible to compare mathematics and science curricula across a wide range of countries, and to serve as the foundation for the development of achievement items for the student assessment part of the study. These frameworks provide a common set of categories that may be used to compare and contrast curricula, textbooks, curriculum guides, or—in the case of this project—examinations.

The TIMSS frameworks were developed for use across all levels of primary and secondary education in mathematics and science. As a result, the number of specific content categories in the frameworks for any given grade or level is quite limited. For this project, the content categories were increased to provide additional detail to the Population 3 level and, therefore, to provide more information about the specifics of curricular similarities and differences across countries. The adapted frameworks are included in Appendix A.

Examining the Examinations complements an analysis of textbooks and curriculum guides that is being conducted in all of the TIMSS countries. In a massive, unprecedented effort, researchers in each of the TIMSS countries are inspecting *every page* of textbooks and curriculum guides for grades 4, 8, and 12. Using methods similar to those described for this examinations study, TIMSS researchers are categorizing the documents' contents and expectations for student performance. Further, the instructional features of the textbooks and guides are being described—including, for example, what proportions of textbooks contain narrative, graphics, exercises or problems, hands-on activities, etc. In combination with the findings from the achievement survey and an investigation of teaching practices, the TIMSS curriculum analysis will constitute a comprehensive source of data that can be used to improve mathematics and science curricula around the world. Reports on the TIMSS curriculum analysis are slated for release during 1996, and reports on student achievement and instructional practices will be published in 1996 and beyond.

The overall objectives of TIMSS are to add to the store of knowledge about what works in the teaching and learning of mathematics and science, and to enable national education systems and policymakers to learn from one another through valid comparisons based on comparable data sets. Formal, highstakes, end-of-secondary school examinations are one means that many countries employ to operationalize important aspects of their curriculum: as intended, as implemented, and as attained. Analyses of such examinations can provide important insights into how countries differ—not only in the topics they deem important enough to include on such examinations, but also in the kinds of performance behaviors students are expected to demonstrate.

Approach to Comparing Examinations[5]

As noted, the results reported in this book speak about only two generally consecutive, years of examinations: usually 1991 and 1992. For most examination features, we have no reason to believe that serious differences existed in other years, but we did not have the resources to pursue that question. If this study limitation is not kept in mind when considering examination topics, inappropriate generalizations could be made. This volume reports topic coverage only over two years of examinations and does not imply that a given country always includes or neglects any particular topic over a longer time. There is good reason to believe that topics change from year to year: in fact, topic differences often occurred between the two years included in the study. Nevertheless, the presence or absence of topics over two years (or even in a single year) can be quite interesting. For example, could the omission of a core topic in a scientific discipline for two years decrease the attention it receives in the classroom?

Translations

The foreign language examinations were translated by professional translators and university personnel. To ensure that the technical nature of the advanced science and mathematics information was preserved, the most technical portions of the examinations were translated by both sets of experts and compared for accuracy.

Overview of Examinations

Researchers reviewed each examination as a whole and wrote a brief profile of it to answer the following questions:

---

[5] For technical details, see Appendix B.

- How many questions are in the examination, and is the examination separated into different major parts?

- How much time are students allowed to spend on the examination as a whole and on each of its parts?

- How many questions are there of each major item type, i.e., multiple-choice, free-response, or hands-on experiment?

- How much choice do students have in selecting questions to answer?

- How can the nature of the examination's topics be characterized at the most general level?

Item-by-Item Comparisons

The researchers next studied every question of every examination, categorizing each for three characteristics—item type, content, and student performance expectations. These detailed analyses were a sizable undertaking: For one thing, in order to describe the questions in these ways, researchers had to think through the solution to each item if not actually answer each question.

Researchers identified the examinations' "scorable events" as the unit of analysis. When compiling results, researchers weighted the values of scorable events by the numbers of associated points indicated by the examinations or their scoring guides, if available.[6] For example, when an examination item had subquestions worth 2, 3, and 5 points, respectively, they were considered to be three scorable events which were weighted by their number of points.

Item Characteristics. Researchers identified each scorable event as multiple-choice, free-response, or a laboratory practical. As pointed out in chapter 2, researchers also noted that several structures were used for multiple-choice items. Similarly, while all free-response questions were open-ended—that is, students had to construct a response rather than choose from ones provided (as in multiple-choice questions)—the length of responses could vary. Researchers noted whether questions required short or extended answers and, in the extreme, whether the answers required were an essay or merely a word or phrase.

---

[6] Points were not available for examinations from Baden-Württemberg and Japan, and for the 1991 French examinations.

Topics. To describe the examinations' topics, researchers augmented the topic lists in the TIMSS Curriculum Frameworks (Robitaille et al., 1993). The resulting set of 50 to 150 detailed topics per subject are intended to describe the content of secondary courses for college-bound students in biology, chemistry, physics, and mathematics subjects. Table 1-2 illustrates the TIMSS content categories and researchers' additions. A complete list of topics used in the study appears in Appendix A.

Table 1-2.   Augmented TIMSS Content Categories: Example of Different Levels

TIMSS, First level topics

1.1	Earth sciences
1.2	Life sciences *(see below)*
1.3	Physical sciences
1.4	Science, technology, and mathematics
1.5	History of science and technology
1.6	Environmental and resource issues

TIMSS, Second level topics

1.2.1	Diversity, organization, structure of living things
1.2.2	Life processes and systems enabling life functions
1.2.3	Life spirals, genetics, continuity, diversity
1.2.4	Interactions of living things *(see below)*
1.2.5	Human biology and health

TIMSS, Third level topics

1.2.4.1	Biomes and ecosystems
1.2.4.2	Habitats and niches
1.2.4.3	Interdependence of life *(see below)*
1.2.4.4	Animal behavior

Fourth level topics, added by researcher

1.2.4.3.1	Food chain webs
1.2.4.3.2	Adaptations to habitat conditions
1.2.4.3.3	Competition among organisms
1.2.4.3.4	Symbiosis, commensalism, parasitism
1.2.4.3.5	Man's impact on the environment

Researchers described the topic of each scorable event, generally as a single topic. But researchers used two categories for description whenever using a single topic would have seriously failed to describe the scorable event. When compiling the examinations' topics, researchers often aggregated the very specific topics found into broader ones for more concise reporting.

Performance Expectations. Researchers used the Performance Expectations of the TIMSS Curriculum Frameworks to describe the performance behaviors required of students in answering the examination questions. The TIMSS student performance expectations are similar to other schema created to describe cognitive operations (e.g., Bloom's well-known taxonomy and many others; see Bloom, 1956); however, the TIMSS schema are more detailed than most others. Table 1-3 summarizes TIMSS Performance Expectations; detailed categories are listed in Table 2-2.

Table 1-3.   TIMSS Performance Expectations: General Categories

Science	Mathematics
Understanding	Knowing
Theorizing, Analyzing, and Solving Problems	Using Routine Procedures
Using Tools, Routine Procedures, and Scientific Processes	Investigating and Problem Solving
Investigating the Natural World	Mathematical Reasoning

Approach to Comparing Examination Systems

Although the primary emphasis of this study was to compare the examinations themselves, considerable effort was also invested in understanding the development and use of the respective examinations. This research provided an important context for comparing examinations and examination systems across countries. We first synthesized the extensive work already done by researchers in comparative education and other fields. Second, because few studies have looked at examination systems with a specific focus on mathematics and science examinations, researchers at the National Center for Improving Science Education and the American Federation of Teachers interviewed dozens of experts to gather contextual information about the examinations. These experts included foreign contacts involved in TIMSS;

officials in foreign education ministries, educational attachés at Washington, DC, embassies; and researchers in the fields of comparative education, mathematics education, and science education.

Recent works particularly relevant to this study are described below. While the volumes by Black (1992) and Eckstein and Noah (1993) describe physics and mathematics examinations, respectively, the other references deal either with examinations in subjects other than science and mathematics or examination systems in general.

- Black (1992) edited a collection of reports about university entrance examinations in physics in France, Japan, the United Kingdom, Sweden, the United States, and six other countries. Each report was written by a representative from that country and includes information about who takes the examinations, how they are written and graded, and how they fit into the educational system; sample questions are also included.

- Eckstein and Noah (1993) take a comprehensive look at the examination systems in China, England and Wales, France, Germany, Japan, Russia, Sweden, and the United States. Their work includes extended vignettes for each country that trace two students' experiences as they progress through the examination process. An appendix contains full language/literature and mathematics examinations from the various countries.

- Eckstein and Noah (1992) edited a collection of essays by various authors about the issues surrounding secondary school-leaving examinations in many countries, including Japan and Sweden.

- Heyneman and Fägerlind (1988) edited a series of essays about university entrance examinations in various countries, including Japan, England, the United States, Sweden, Australia, and China. The book also has essays discussing comparisons of such examinations and the role of examinations in assessment.

- In 1991, the National Endowment for the Humanities released a collection of advanced-level humanities examinations from France, Germany, England and Wales, and Japan. Each set of examinations is preceded by a short essay describing the examination system in that country.[7]

---

[7] This publication spurred wide interest among the media, as well as within the educational and policy communities, as a means of obtaining subject-specific understanding of other nations' examinations.

- For the US Office of Technology Assessment, Madaus and Kellaghan (1991) reported on European examination systems and what the United States can learn from them.

Overview of Examination Systems

The contrasts among the internal features of countries' examinations can best be understood in the context of national differences in examination purposes, courses of study, etc. For quick reference, the following overviews provide key facts about the seven examination systems studied here. Part II of this book is devoted to more extensive consideration of this, providing detailed descriptions of examination systems and comparisons among them.

England and Wales

At age 16, students in England and Wales end their compulsory education by taking a series of subject-specific examinations to earn a General Certificate of Secondary Education (GCSE). Students may then elect to continue their academic schooling for two years in preparation for the Advanced Level examinations (A-levels) required for college admittance.

In secondary school, all students study a large number of subjects set forth by the national curriculum, of which five must be studied through age 16. After receiving their GCSE, however, students intending to go to college usually concentrate on three subjects and spend more than 70 percent of their time on these (Kelly, 1994). At the end of the second year, these students take A-level examinations in each subject they chose to study. Most colleges in England and Wales require applicants to pass either two or three A-level examinations; higher grades increase candidates' chance of acceptance. In 1992, 31 percent of the age cohort took A-level examinations in at least one subject, and 25 percent of the cohort earned passing grades (Gandal et al., 1994).

Seven private organizations, called examining boards, create and administer the A-level examinations in England and Wales. These boards are overseen by a national governmental body, the School Curriculum and Assessment Authority. Although each school may elect to use examinations from any board, boards sometimes have dominance in a geographical region. The examinations are written and scored by committees of secondary school teachers and university subject specialists. The two years of examinations described in this book came from two examination boards: the University of London Examinations and Assessment Council, and the Associated Examining Board.

France

Students' course grades and performance on brevet examinations at age 15 determine, for those students continuing schooling, whether they will proceed along an academic, vocational, or technical course of study. To be eligible for university admission, students in France must earn a baccalauréat diploma by passing a series of examinations in the subject areas required while studying at a lycée (the equivalent of grades 10 through 12). Passing the baccalauréat does not guarantee placement into all university subjects. The most popular subjects may require minimum scores above a simple pass for entrance (Colomb, 1995).

The subject examinations a student takes, and the weight of each examination toward the final composite score, differ depending on which track the student chooses to complete. Regardless of the track selected, students receive a broad education encompassing many disciplines and are expected to demonstrate varying levels of mastery in many of these subject areas. For instance, students in Track C—an academic track that emphasized the study of mathematics, physics, and chemistry—also had to take written examinations in French, history and geography, biology, and philosophy, as well as an oral foreign language examination. In 1992, 51 percent of the age cohort earned either an academic baccalauréat or one of the two types of vocational baccalauréats (Langlois, 1992). Forty-three percent of the age cohort tried for an academic baccalauréat, and 32 percent earned one (Bodin, 1994).

Twenty-six regional academies grouped into four clusters—two of which are the Paris and Aix regional clusters discussed in this book—develop and administer baccalauréat examinations under the direction of the national education ministry. The examinations are written and scored by teams of secondary school teachers.

Germany

Parents and teachers usually decide which of three tracks students will pursue after they have completed primary school at age 10: Hauptschule, Realschule, or Gymnasium. Each track has its own schools and curriculum. The curriculum for university track students, in Gymnasium, aims for both breadth and depth of knowledge; it culminates with the Abitur certificate. This certificate is awarded based on a combination of student grades over the final two years of secondary school course work and scores on a series of four Abitur examinations. Admittance to the most popular university subjects requires not only an Abitur, but also a minimum-level Abitur scores (Führ, 1989).

During their last two years of Gymnasium (equivalent to 12th and 13th grades), university track students complete 28 courses—22 at the basic level and 6 at the advanced. They also select four subject areas in which they will eventually take an Abitur examination. To ensure that all Abitur students reach proficiency in a broad range of subjects, examinations must be taken in each of the following three fields of knowledge: (1) language, literature, and the arts; (2) social sciences; (3) mathematics, sciences, and technology. Students will take two examinations in one of the latter fields. To ensure depth of knowledge, at least two of the four examinations must be taken at the advanced level. In 1991, 37 percent of the age cohort took the Abitur, and 35 percent of the age cohort passed (Gandal et al., 1994).

In a majority of German Länder, part of the teacher corps's professional responsibility is to develop and score examinations following guidelines of the state and national education ministries. In a few Länder, the state ministry develops the Abitur. The two years of examinations analyzed in this volume come from two such Länder: Baden-Württemberg and Bavaria.

Israel

At about age 15, students may enter the academic track which eventually leads to the Bagrut certification needed for university admission. Students are not tracked until this point in their education, except in English and mathematics. They receive the Bagrut by achieving satisfactory grades overall, composed of an average of their course work and external examination scores (Tamir, 1993).

In general, students have a great deal of flexibility in planning their course of study. Each course and accompanying examination is worth from 2 to 5 points, depending on the level. In all, students must accumulate 22 points, only half of which are for compulsory subjects (English, mathematics, Hebrew, and the Bible). Although receiving the Bagrut guarantees admission to university, it does not guarantee admission into all subjects. The most popular subjects base admission on Bagrut examination score and additional examinations. About 40 percent of the cohort receives the Bagrut (Tamir, 1994).

The examinations are written by committees appointed by a chief inspector for each subject; inspectors in turn are appointed by the Minister of Education. The examinations follow the syllabus and examination guidelines established by the appropriate federally appointed subject instruction committee. The examinations are graded by experienced teachers during the summer. Teachers do not grade their own students' papers. The examinations analyzed for this report are advanced (5-point) science and mathematics examinations.

Japan

Japanese students take competitive examinations at the end of lower secondary school that determine which high school they will attend. Consequently, while all high schools' courses of study follow a national curriculum, the caliber of the student body varies among schools as does the distribution of students not attending college or attending colleges of various rankings in quality. The best Japanese high school students progress through two tiers of college entrance examinations to be eligible to apply for university admission at all national and local public universities and some private colleges. High school seniors first take the University Entrance Center Examination (UECE), which is based on the national curriculum. National public universities require students to sit for the UECE, but most private universities do not. In Japan, the private universities are generally held to be less prestigious than the public universities, although this perception appears to be changing.

All prospective college students must take sets of entrance examinations developed and administered by the individual universities to which they apply; because of scheduling, they cannot take examinations from more than two public institutions. Competition for admission is very intense, and many students who are not accepted to the school of their choice become ronin, "masterless wandering samurai," who spend one or more years studying to retake the entrance examinations. In 1990, 43 percent of high school students graduating that year applied to four-year universities, and thus took individual university entrance examinations (Ryu, 1992). In addition, many high school graduates from previous years (nearly one-third of 1990 applicants) applied to universities; many were second- or third-time applicants. Sixty-three percent of 1990 applicants were admitted to university or college that year (Ryu, 1992).

At the top of the hierarchy of Japanese universities is Tokyo University, whose entrance standards are widely considered to be the most difficult. This book discusses the features of two years of examinations from Tokyo University.

Sweden

During ninth grade, students select the course of study, or line, they would like to follow in upper secondary school; grades are used as the criteria for admission into the more popular lines. Although it is noncompulsory, 90 percent of students finish upper secondary school. Admission to university for students graduating from upper secondary school is based on their grades. These grades, in turn, are based on school performance and central examination scores, with the exact ratio independently determined by each classroom teacher.

College-bound students take about 14 courses during their three years of upper secondary school and take external examinations in six or seven subjects (Gisselberg and Johansson, 1992; Eckstein and Noah, 1993). The subjects for each line differ, but all tend to have components in language, science and mathematics, and social studies. Five academic lines are oriented toward postsecondary study: natural science, technology, social sciences, economics, and liberal arts. The science and technology lines both require varying amounts of courses in mathematics, physics, chemistry, biology, foreign languages, arts, and social studies. Some examinations are given at the end of the second year of study, and others at the end of the third and final year. The physics and mathematics examinations discussed in this volume were given at the end of the third year, while chemistry was given in the spring of the second year. About 45 percent of the age group completes one of the academic lines.

The National Agency of Education commissions institutions (such as the University of Umeå Department of Educational Measurement) to prepare these examinations. These institutions prepare the examinations in consultation with a reference group (appointed by the institution) of experts in the field. The examinations are scored and graded by students' classroom teachers. These teachers may decide how to combine the examination score with their own assessments of the students in determining the final grades, but the distribution of the final grades as a class must match the distribution of the class's examination scores.

United States

Although there are not particular curriculum or examination standards that all college-bound students in the United States must meet, there are several examinations that are used by college admissions offices in making their selections. The best known and most widely taken examinations are the Scholastic Achievement Test and the American College Testing Assessment (ACT), neither of which is tied to curricula taught in schools.

The Advanced Placement examinations are the only curriculum-based assessments taken by US college-bound students. Students who choose to participate in the AP program take AP courses in any of nearly 20 subjects. After completing the courses, students may elect to take the AP examinations, which cost $67 each in the spring of 1995. These examinations require a greater knowledge of subject matter than the SAT I or the ACT. In 1993, 6.6 percent of 18-year-olds took one or more AP examinations; 4.3 percent passed them (AP Program, 1994).

Most college and universities will factor AP examination scores (or grades in AP courses taken in the senior year) in their admissions decisions. Although each admitting institution has its own policy for weighing these examinations in its admissions criteria, students with high scores or grades are generally considered to have an advantage over other applicants who have not taken advanced course work. Additionally, most institutions award students with high scores with some college credit in certain subjects.

Part I

*Looking Inside
the Examinations*

Paul Black

THE IMPORTANCE OF THIS study is that it contributes powerfully to a necessary sophistication in our understanding of international comparisons in education. League tables produced by listing scores on a common test of pupils from many countries inevitably attract a great deal of attention. They have been known to spur politicians into criticism and even action, while they often put teachers and schools on the defensive. This is not merely because of the inevitability of half the countries being below the median, but also because, like most statistics, what they conceal is as important as what they reveal. This is because pupils' performances on a particular question depend very strongly on the extent to which its demands are familiar to the pupil and on the opportunities the pupil has had to learn about responding to such demands. Thus, the effectiveness of the teaching or the commitment of the pupils are only two of several important determinants of test outcomes. The curricula and the inter-related practices of teaching, learning, and testing to which pupils are accustomed are of equal, arguably greater, importance. So comparisons of these are essential.

What this study shows is the enormous variety between countries, both in the nature and range of topics assessed, and in the ways in which their national assessments are carried out. This is both daunting and comforting. It is daunting to see how differently the different national traditions go about the same task. This raises questions—have many countries got it wrong, or is best practice so entwined in the culture that comparisons are of little value? Both of these are valid questions, and one could agree in part with all of them in the interpretations that they imply. Of course, another interpretation is that it doesn't much matter how assessment is carried out: this we must reject, at least as a working rule, until all other possibilities have been explored.

The variety is comforting in that it shows that one's own national or regional system does not have the inevitability that custom and practice seem to imply. It is possible to assess practical work, it is possible to use more thought-provoking questions, if some countries are able to spend longer on assessments maybe the price of this is worth paying, and so on. Thus, the comparisons can be a spur to improved practice, to lay down criteria for worthwhile improvements.

Nevertheless, the comparisons do raise serious questions about the quality of testing and assessment. One that stands out here must be about the time taken for external tests. If one country invests over eight hours in tests in one subject to determine future career prospects, how can another perform the same function in under two hours? In both cases, the test instruments could be reliable, in the sense that they produce good indices on checks which affirm their internal homogeneity - although, for some of the tests described here, even this is not evaluated. However, such a check is by no means adequate. Any test is meant to sample a domain of performance, and where such domains have been set out comprehensively and explored for school performance in science and mathematics, it has been clear that many hours of testing would be needed to assess with any accuracy. One way to overcome the difficulty is to narrow the range of the domain assessed, and this effect is evident in many of the systems described here.

Here the issue of reliability overlaps with that of validity, for a narrowing of the domain secures better reliability at the expense of narrowing the scope of subject aims which the examinations exhibit and thereby, because of their high stakes in most systems, the aims that they impose on school learning. Here it is evident in all four subject areas that there is serious cause for concern, in two main areas.

One is in the balance between questions which can be answered by routine procedures using learned algorithms and those which require thoughtful translation and application of principles and procedures. In most systems, the analysis shows that the balance is very much in favor of the former, and this must in part be due to the limitations of test times. However, only in part - where a test can be highly selective and does not have to concern itself with doing justice to the average pupil, then open ended and complex tasks can be used - the University of Tokyo is a notable example here. However, the very difficult task for most test setters is to develop questions which demand thoughtfulness rather than rote learning, at a level where the average pupils can have a good chance of success. There are too few examples of success here, and we all need to look for good examples to help such development.

The second main area is in practical work in science and mathematics. This is ignored in many systems, and treated in others by paper and pencil surrogates rather than by actual work with equipment. Where this issue has been researched, it comes out that the surrogates are not effective - pupils behave quite differently in the two types of assessment. However, some of the practical tests reviewed here are not encouraging either - the limitations of a short timed test often lead to exercises which are more about obeying instructions than about doing genuine science.

Thus it would seem that the only way out is to assess work that is done in more relaxed conditions and over a period of time, which means that the assessment has to be in the hands of teachers. Teachers are trusted to assess in some countries and not in others. Where teachers are trusted, there is further range of practice in the amount of effort devoted to setting up external guidelines and procedures, both to calibrate standards between schools and to improve the assessment skills of teachers. Some experience has been gained in doing this in the UK and the USA, but unfortunately the examples selected here do not include such cases - partly for the good reason that they are uncommon. However, what is here shows powerfully that there is a need for work in this direction to be explored.

To open up such issues is to open a Pandora's box. If we are to be ambitious to establish testing regimes which help, rather than harm, valid and thoughtful learning, if we are to be ambitious that preparation of examinations can give the pupil an authentic and attractive experience of what it can be like to do mathematics and science, then the task expands in its implications as it is explored in greater detail. Change to radically improve assessment can require changes in curricula, in teacher training, and in the nature and cost of national systems. Yet if examinations are as important to pupils and to the future of the subjects as they seem to be, then such change should be vigorously explored. The great virtue of this study is that it opens up possibilities, calls in question the inevitability of existing practices, and exposes some of the dysfunctional aspects which should serve as a spur to thought and action.

2

Comparing Examinations Across Subjects and Countries

Edward Britton
John Dossey
Dwaine Eubanks
Lucy Eubanks
Kjell Gisselberg
Simon Hawkins
Senta Raizen
Pinchas Tamir

OUR ANALYSIS OF INTERNATIONAL examinations in biology, chemistry, physics, and mathematics covered approximately 5,000 questions and subquestions in 77 examinations from seven countries, as listed in Table 1-1. This chapter contrasts the internal features of these examinations, summarizing patterns derived from the full-scale, detailed analyses in Chapters 3 through 6.

In general, the differences among the examinations outnumber their commonalities, but the nature and degree of these differences vary for each examination feature. Obviously, there are many ways in which to present the variations so as to best contrast examination features. The contrasts depend on how examinations are grouped for comparison—by or within country, or by subject. For example, chemistry examinations could be compared with other subject examinations, across all seven countries; French examinations could be compared with other countries—examinations, regardless of subject; or examinations from different examination boards within England and Wales could be compared. What we have done in this book—and what we summarize in this chapter—is a comparison of the examinations in each of these ways: by country, by subject, and within country. We also examined variations from year to year within each country's examinations but, with rare exceptions, found the differences to be very minor; hence results are aggregated across all years analyzed for each examination.

Chapter 1 provides background information for this volume, including an overview of study methods; they are elaborated further in the technical notes of Appendix B. This chapter opens with a brief section highlighting noteworthy contrasts among the examinations. It then presents concise summaries of examination differences for the following topics drawn from the major sections of Chapters 3-6:

- general structure (examination components, length, main item types, student choices among questions);
- item characteristics (detailed item types and use of diagrams, graphs, and tables);
- examination topics; and
- performance expectations (student operations required by examination questions).

The chapter then presents conclusions about these findings in terms of, first, the relative difficulty of the examinations and, second, the curricular implications of the analysis.

Highlights

The following highlights summarize the contrasts among the examinations analyzed in this study. We include the most notable findings from the perspective of science and mathematics education. The highlights are grouped by analysis topic; for more information, including implications, the reader should refer to the corresponding chapter sections.

General Structure

- Most examinations were 3 to 4 hours long (France, Germany, Sweden, and the United States); however, English/Welsh examinations were much longer (average 7.25 hours), Israeli examinations were notably longer (average 5.25 hours), and Tokyo University examinations were slightly shorter (2.5 hours).

- Examination length was quite consistent within each country, except among the examinations in England and Wales. There, examinations in different subjects varied by up to 2 hours, and examinations from different examination boards varied by up to 3 hours.

- A substantial portion of each examination in England/Wales and Israel gave students extensive choice among questions; there was choice for only one subject in France and the United States. In contrast, all questions in Germany, Japan, and Sweden were compulsory.

Item Characteristics

- The US Advanced Placement (AP) examinations in every subject allotted half the examination time to multiple-choice items. The multiple-choice portions of examinations in England/Wales, Israel, Japan, and Sweden were found only in some subjects, and represented a much smaller proportion of the examination. France and Germany did not use multiple-choice items.

- All examinations made extensive use of free-response items. These items accounted for every question in France and Germany; the great majority of questions in England/Wales, Israel, Japan, and Sweden; and 40 to 55 percent of the Advanced Placement examinations.

- Laboratory practicals were found only in England/Wales (mostly in chemistry and physics) and Israel (biology and physics).

- Required essay questions (a single free-response item allotted at least 20 minutes) were found only in one or two subject examinations in England/ Wales, France, Israel, and the United States.

- Only an average of about 15 percent of each country's biology and physics examinations and 5 percent of their chemistry examinations incorporated tables and graphs.

Examination Topics

- Some core topics in the field of biology were covered universally (e.g., cells), but the examinations generally gave little or no attention to some other important biology topics (e.g., evolution).

- All chemistry examinations except the AP devoted considerable attention to organic chemistry and some attention to industrial applications, but no countries included environmental chemistry.

- Physics examinations differed in their emphasis on classical versus modern physics. They rarely cast phenomena in real-world contexts.

- All mathematics examinations treated the subject as an abstract discipline, with no evidence of modeling, attempts to connect mathematics to real-world problems, or using other disciplines as contexts.

Performance Expectations

- Only 5 to 25 percent of the science examination scores in each country arose from expectations that students perform tasks inherent to scientific or mathematical experimentation: for example, designing or conducting investigations, interpreting data, and using routine procedures. The countries with the greatest emphasis on these tasks were England/Wales, France, and Israel.

- Conversely, in each country, 75 to 95 percent of the science examination score was allotted to questions requiring students to work with scientific concepts by means of these tasks: showing understanding of concepts, and using concepts to solve quantitative problems or explain scientific phenomena.

- Except in Japan, little of the mathematics examinations required mathematical reasoning. Rather, they focused on using routine procedures and problem solving.

Difficulty

- A country's examinations could be considered either more or less difficult than those of other countries, depending on which of the many specific examination characteristics is being considered. A single ranking of examinations or of countries for difficulty is an oversimplification.

Structure of Examinations

The structure of the examinations varied considerably among countries by length, item type, and the amount and kinds of choice students had among questions. For example, some examinations were hours longer than others. And, while every examination included free-response questions, only some made use of multiple-choice questions, and very few incorporated laboratory practicals. Some countries let students choose among questions for substantial portions of their examinations; in other countries, all examination questions were compulsory.

Examination structure varied within some countries as well as across all countries. Marked differences existed among the structures of the various examinations given in England and Wales; this was true to a lesser extent within Israel, Japan, and Sweden. On the other hand, the various examinations within France, Germany, and the United States were fairly uniform in structure.

Figure 2-1 portrays, across all four subjects, the general composite structure of examinations in each country except England and Wales, where the differences across subjects and examination boards were too great to permit development of a country composite. The figure shows (1) how many hours students were given for taking the examinations and (2) the amount of time that examinations devoted to major item types (multiple-choice, free-response, and laboratory practicals).

For comparable information by subject, see Figures 3-1, 4-1, 5-1, and 6-1 in Chapters 3, 4, 5, and 6. For comparable information for England and Wales, see Figure 2-2.

Overview of Examination Structure

England and Wales. At 6 to 9 hours in length, these were the longest examinations. Most of the science examinations contained laboratory practicals. In general, all of the examinations relied primarily on free-response questions, but some also employed multiple-choice items. Examinations generally had four separately administered (timed) "papers," one of which was the practical. Students had choices among questions in several papers of each examination.

France, Germany, and Sweden. These examinations ranged in length from 3 to 4.5 hours. The most distinctive characteristic of the French and German examinations is that all of the questions required students to construct their answers. This was also true for the Swedish examinations, except in chemistry which contained a few multiple-choice items. All questions on these three countries' examinations were mandatory, except for one choice between sections of the French biology examinations.

Israel. The Israeli examinations were between 5 and 5.5 hours long. The biology and physics examinations included a laboratory practical, and all examinations used a variety of multiple-choice and free-response items. Examinations comprised two or three separately administered papers. Students were given considerable choice among questions in some papers.

Figure 2-1.   General Structure of Examinations in Seven Countries
Average Length and Item Type Amounts for all Subjects

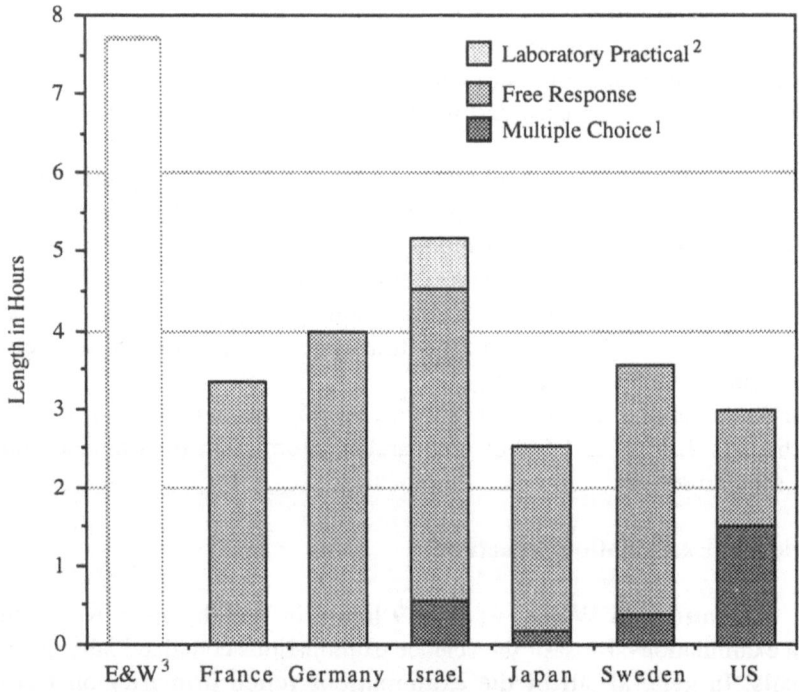

1. Multiple Choice Items —
   Israel: multiple choice in biology, chemistry only.
   Japan: multiple choice in biology, chemistry only.
2. Laboratory Practical — Israel: practicals in biology and physics only.
3. Because examination in England/Wales vary greatly among different subjects
   and examination boards, results are reported separately in Figure 2-2.

Japan. At 2.5 hours long, the Japanese examinations were the short-est, but only by about 30 minutes. Free-response questions dominated, but a few multiple-choice items were included. All questions were mandatory.

United States. These 3-hour examinations had two separately timed sections, one consisting of more than 100 multiple-choice questions and the other of a few free-response questions. Only the free-response section of the chemistry examinations gave students choices among the questions to be answered.

Figure 2-2.   General Structure of Examinations in England and Wales
Lengths and Item Type Amounts in Each Subject

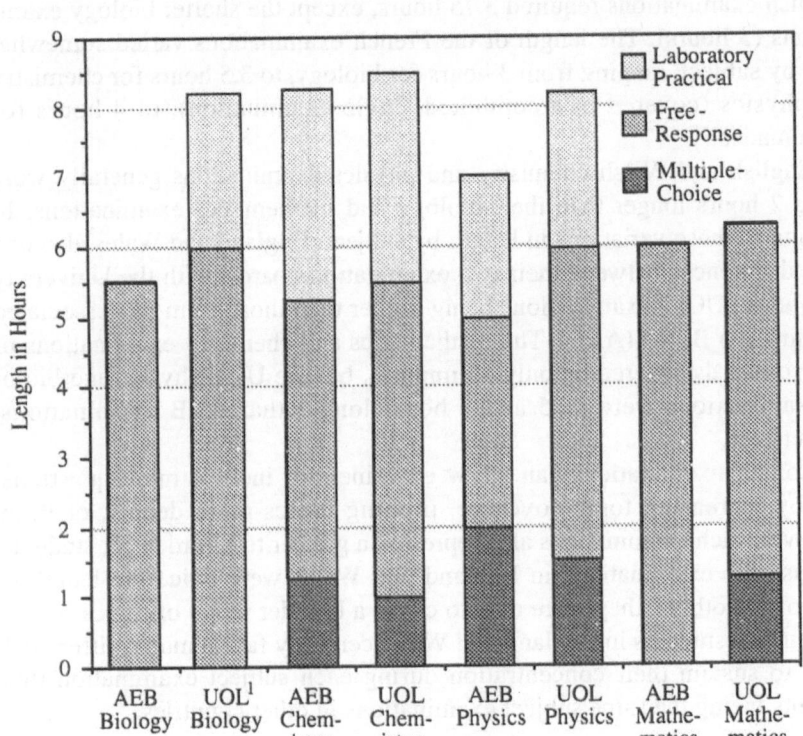

1. This represents 1991 only, the 1992 UOL biology examination had no laboratory practical, and lasted nine hours.

Examination Length

England and Wales had by far the longest examinations. Their average length of 7.25 hours[1] is 2 hours greater than the next-longest examinations (Israel, 5.25 hours). Most examinations were 3 to 4 hours long (France, Germany, Sweden, and the United States). Tokyo University examinations were a little shorter, averaging 2.5 hours.

Within each country except England and Wales, examinations in biology, chemistry, physics, and mathematics often were the same length and never differed by more than an hour. In Japan, all the examinations were 2.5 hours;

---

[1] Because all examination lengths were in multiples of one-quarter hour, average lengths are reported to the nearest 0.25 hours.

in the United States, they all were 3 hours. The German examinations took 4 hours, except for the slightly longer biology examinations (4.5 hours). All Swedish examinations required 3.75 hours, except the shorter biology examinations (3 hours). The length of the French examinations varied somewhat more by subject, ranging from 3 hours for biology, to 3.5 hours for chemistry and physics (covered in a combined single examination), to 4 hours for mathematics.

English and Welsh chemistry and physics examinations generally were 1.5 to 2 hours longer than their biology and mathematics examinations. In addition to these variations in length by subject, England and Wales also had some differences between their two examination boards, with the University of London (UOL) examinations being longer than those from the Associated Examination Board (AEB). The mathematics and chemistry examinations of the two boards differed by only 15 minutes, but the UOL physics and biology examinations were 1.25 and 3 hours longer than AEB examinations, respectively.

Longer examinations can allow examiners to include more questions, thereby increasing topic coverage, probing topics more deeply, or both. However, such examinations also represent a greater test burden for students. Because the examinations in England and Wales were twice the length of most of the others, they were able to cover a broader range of topics. As for test burden, students in England and Wales certainly face a much stiffer challenge to sustain their concentration during each subject examination than students facing the same subject examinations in other countries.

Main Item Types

All countries used free-response questions extensively for all subjects, and many used some multiple-choice items, but only two had students do laboratory practicals. (See Figures 2-1 and 2-2.) All examinations save two grouped items of the same type together. The exceptions were the Japanese and Swedish examinations, which distributed their item types throughout. Since various item types generally require different cognitive processes, Japanese and Swedish students had to alternate their thinking approaches more often during the course of an examination than did students in other countries. This was not a frequent occurrence, however, since these two countries' examinations only employed a few multiple-choice questions.

Free-response items may require only a word or phrase answer, or may require an essay, but they are all open-ended—that is, students must construct a response rather than choose from ones provided (as in multiple-choice questions). The French and German examinations were entirely composed of free-response questions; this item type dominates the examinations in other countries as well. The AP examinations made the least use of this item type, allocating only half of total examination time to free-response questions.

Only AP examinations included a major multiple-choice component: this item type accounted for half of the total AP examination time in every subject. Multiple-choice questions represented a much smaller proportion of total examination time in England/Wales, Israel, Japan, and Sweden. Further, in these countries, only examinations in some subjects made use of multiple-choice items.

No countries except England/Wales and Israel included laboratory practicals in their examinations. Both years of AEB and UOL examinations used practicals in chemistry and physics; UOL also included a laboratory in its 1991 biology examination. In Israel, only biology and physics examinations incorporated laboratory practicals.

Choice

All examinations in England/Wales and Israel gave students some choice among questions. In contrast, choice occurred in only one subject each in France and the United States. All questions in Germany, Japan, and Sweden were compulsory. (See Table 2-1.)

Table 2-1.   Student Choice Among Examination Questions

Examinations with choice in all subjects	Examinations with choice in one subject	Examinations with no choice
England & Wales Israel[1]	France (biology only) US (chemistry only)	Germany Japan Sweden

[1]   Choice found in biology, chemistry, and physics; mathematics examinations not available for analysis.

Overall, the amount of choice provided was substantial, and was presented in quite different ways. The French biology examinations essentially offered students two different examinations by letting them choose one of two parts. In the United States and England/Wales, all multiple-choice portions of examinations were compulsory, but their free-response sections built in some interesting choice structures. The AP chemistry examination, for example, had several combinations of choice: select one of two problems, three of five, and five of eight. The English and Welsh examinations had similar choices, such as selecting one of two questions or two of five.

Given the direct effect that high-stakes examinations have on secondary curricula, including more topics in the examinations increases the likelihood that day-to-day instruction will have topic balance. But examination time limits the number of questions students can answer. Examination designers can incorporate more topics by permitting choices among questions, thus adding more questions and topics. It is interesting that the examinations with the least time constraints were the very ones that offered the most choice. English and Welsh examinations offered many small choices despite being 2 to 3 hours longer than other examinations.

The constraint that examination time places on topic coverage is accentuated when examinations include substantial numbers of free-response items. Because they often are very time consuming, only a few free-response questions can be included in an examination. Yet while German, Japanese, Swedish, and most French examinations depended heavily on free-response items, they did not use choice to broaden topic coverage.

Another effect of student choice is reduction in test burden. Students have the opportunity to select topics they think they know and avoid others. Students experiencing examinations with choices may perceive them to be less threatening. Whether they actually pick to their strengths and consequently increase their examination result is not clear. Thissen, Wainer and Wang (1994), after conducting a statistical analysis of students—choices and results on recent AP examinations (including one of the AP chemistry examinations analyzed for this book), raise doubts about the wisdom of offering choices. They found mismatches between the topic strengths indicated by students—scores on multiple-choice questions and students—topic choices among free-response questions on the same examinations:

> Men had an advantage: not because of any superior knowledge of chemistry, but merely because they tended to choose easier items to answer . . . Examinees, when choosing an item, do not know their probability of responding correctly to the item. Instead, they have some subjective ideas of that probability. [There is] strong evidence indicating that some examinees do not choose wisely (pp. 165, 170).

While the study was limited to AP students and examinations, and these may not be characteristic of students and examinations in other countries, the study findings at least raise questions about the effects of choice in high-stakes examinations.

The Israeli examinations had considerable choice, but it mirrors topic choices made in the students' last two years of study rather than being a testing device to increase topic coverage or reduce test burden. For example, biology students and their teachers choose six of nine "basic" topics and three of six "in-depth" topics to study during grades 11 and 12. (See Chapter 3 for details.) When taking examinations, students choose the topics they have studied. Hence, the examinations are aligned with a curriculum designed to permit more in-depth learning of fewer topics.

Item Characteristics

This section begins by contrasting variations among examinations in their amounts of multiple-choice, free-response and laboratory items. While Figures 2-2 and 2-3 reported the relative amounts of major items types in examinations of different countries, this section compares the occurrence of each item type in more detail and describes the variance by subject as well as country. More specific item types are discussed. For example, free-response items are differentiated by the length of answer they require students to provide—a word or phrase, short or extended answers, or essays.

In addition to describing differences in the amounts of each item type, this section elaborates their characteristics—for example, the numbers and configurations of response options in multiple-choice questions or the kinds of interdependence built into linked free-response questions. We end with a look at the amounts and types of photographs, diagrams, tables and graphs employed in the examinations.

Multiple Choice Items

Figures 2-1 and 2-2 showed that the amount of multiple-choice items in examinations varied notably across countries, and Figure 2-3 indicates that within countries having multiple-choice items—every country except France and Germany—most used different amounts for biology, chemistry, physics and mathematics. Only the US examinations used similar amounts of multiple-choice items in all subjects. In contrast, 25 percent of the Japanese biology examination score was drawn from multiple-choice questions while Japanese physics and mathematics examinations contained no multiple-choice items and the chemistry examinations had very few. Israel and Sweden also had very disparate use of multiple-choice items in examinations for different subjects.[2]   Use of multiple-choice items was fairly even across different subjects in England and Wales, except that the English/Welsh biology examinations had none. Since an examination with a substantial number of multiple-choice items is different from one that does not have any, students in England and Wales, Israel, Japan and Sweden face quite different examinations in different subjects. Based on this examination feature alone, it certainly cannot be said that national examinations in these countries are standardized.

Figure 2-3.   Use of Multiple-Choice Items: All Examinations

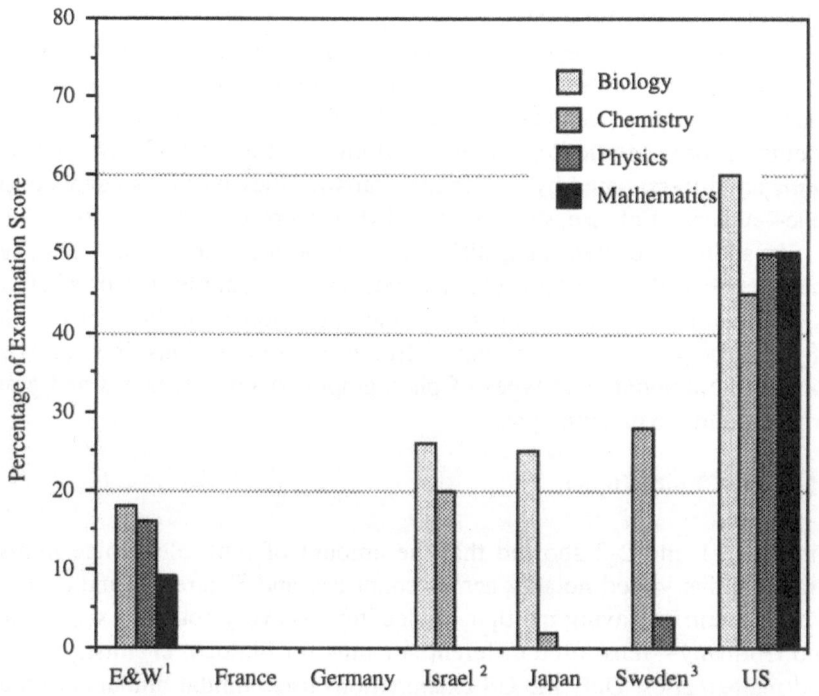

1.   Values average data from AEB and UOL. While their results for chemistry and physics are very similar, the boards' mathematics results differed. (AEB, 9%; UOL 17%)
2.   No mathematics examination was available from Israel.
3.   Sweden has no biology matriculation examination.

[2] The blanks in Figure 2-3 for mathematics in Israel and biology in Sweden are there because no Israeli mathematics examination was available to the study and Sweden does not have a biology examination.

Scientists, mathematicians and educators should note from Figure 2-3 that few generalities can be made about the relative use of multiple-choice items within subjects. All five countries having multiple-choice items used them in chemistry examinations, although marginally so in Japan. However, only three countries used them in biology, a different set of three countries used them in physics, and two countries employed them for mathematics examinations.

The examples of multiple-choice questions provided in chapters 3 to 6 reveal that multiple-choice is not a monolithic item type, although the standard fare was a stem followed by discrete, structured response options. But even this most common of item structures varied. Some items had four options while others had five—a difference implying unequal difficulty since the odds of guessing correctly in the former configuration are greater than in the latter. Different numbers of responses occurred between countries and within England and Wales, where the AEB examinations used four options and UOL examinations used five.

The four or five response options were linked in some items, in ways that made some questions easier and others more difficult. Examinations in England and Wales employed several more complex questions, like Example 6-2, which forced students to consider combinations of responses. Some US and English and Welsh examinations provided the same responses for a set of questions, as in Example 3-1, which can sometimes make it easier for students to narrow their choices among the responses as they proceed through the question set.

Example 3-2, used in the 1991 Japanese biology examination to assess students' understanding of photosynthesis, had a very complex structure—the most elaborate multiple-choice format in the study. Students first had to choose one of three responses and then two of eight other responses. But, they also had to order their set of three responses to match the steps in photosynthesis. We caution that this item does not characterize Japanese examinations since it was the only one of its kind in the biology examination, and Japanese examinations in other subjects did not use multiple-choice items to any extent.

But Example 3-2 nevertheless is instructive; it makes the typical delineations between multiple-choice and free-response items less clear. Since it still is a structured-response item, the question can be readily and objectively scored. But if an examination were to use such lengthy items extensively, a testing advantage that multiple-choice items typically offer—covering a large number of topics in a given time—would be less pronounced or even erased. Such an item certainly takes longer than some free-response questions that only require a short answer. Given the complexity of Example 3-2, it certainly is much harder than typical multiple-choice questions and probably rivals the difficulty of many free-response items. Regardless of its difficulty, however, this item still requires a different cognitive process than free-response items do, i.e., students still were able to choose from provided responses rather than having to construct them.

Free Response Items

We previously noted that free-response items were the most used item type in *every* examination and the only type employed by French and German examinations. As was the case with multiple-choice items, however, all free-response items were not alike. This section contrasts the examinations' use of free-response items by the length of responses that they required students to construct.

Word/Phrase Answers. The briefest answers needed to answer any free-response items were merely a word or phrase. The formats for word/phrase questions included: stating the term that describes a given concept; providing a word or phrase left out of a sentence or paragraph ("fill-in-the-blank"); identifying structures in diagrams; and stating the outcome of a routine experimental procedure. A chemistry question that illustrates the last format is predicting what color an indicator would turn in a specified solution. Word/phrase questions constituted about one-fifth of the Japanese biology examinations, one-tenth of the biology and chemistry examinations in England and Wales, and less than five percent of some other chemistry examinations. These questions were not found in mathematics or physics examinations and were used sparingly in a few biology and chemistry examinations.

Because word/phrase questions are less difficult than free-response items that require longer responses, it was not surprising to find limited use of word/phrase questions in these challenging examinations. Their absence from mathematics examinations is consistent with the field of mathematics where terms, which can lend themselves to word/phrase answers, are less common than in the sciences. But why all physics examinations omitted this item type while some biology and chemistry examinations included them is a puzzle, since all three sciences have an abundance of terms or concepts. Finally, perhaps the relatively high use of word/phrase items in Japan's biology examination and absence in its chemistry and physics examinations tends to make the former easier than the latter. This could partially explain why Japanese students, endeavoring to demonstrate the most advanced achievement possible, would more frequently opt to study chemistry or physics in high school and be tested in these subjects rather than biology. (See Nakayama, 1994, in the Japan section of Chapter 8.)

Short or Extended Answers. The majority of free-response items required short or extended answers, i.e., from one to several sentences of text or quantitative answers requiring one or multiple calculation steps by using one or more equations or formulas.[3] The great variety of questions that corresponded to this definition can be seen by a quick perusal of the many sample questions throughout Chapters 3 through 6. Perhaps one of the most novel items was an Israeli question, Example 3-4, that provided a multiple-choice question with the correct answer already indicated. The students' task was to explain why that answer was the correct one.

The dominance of short- and extended-answer questions among free-response items of these examinations is typical. Much less research has been done on the effects of variations within this format than on multiple-choice formats. There is, however, one structural feature of the examinations' short- and extended-answer questions meriting comment—the nearly universal use of sets of dependent items, i.e., where the answer to one question depends on a previous answer. Stating the now familiar refrain of inconsistency within countries, we generally found the extent to which a given country's examinations incorporated dependent sets of free-response items varied by subject. For example, over 50 percent of free-response items in the Israeli biology examination were dependent upon each other compared to only about 25 percent of the Israeli physics items. Similar amounts and differences were noted for most countries or regions but the US examinations were the exception, having dependent questions for about 40 percent of the free-response items in the examination for each subject. The Swedish examinations used less of this feature than did examinations from other countries.

Examination writers can use question sets to probe different aspects or levels of understanding about a topic as students progress through the set. This question structure also economizes on testing time, since one problem context can be used for several questions whereas an equal number of independent questions would require reading and thinking about a different problem context for each. Since the use of dependent sets varies so much among the examinations in different subjects and countries, perhaps the scoring issues often associated with dependent questions have not been adequately considered. The scoring information available to us often was unclear about the extent to which students would be penalized for answering a given question incorrectly—not because their response to that question was wrong, but rather because they started off with an incorrect answer to a previous ques-

---

[3] Appendix B explains how the authors of Chapters 3 to 6 distinguished between short and extended answers. Because delineations between these two item types were not always comparable across subjects, we discuss the set of examinations here by collectively reporting short- and extended-answer questions.

tion. To the extent such scoring dependence is prevalent, examinations with considerably different amounts of dependent questions present unequal amounts of scoring handicaps to examinees.

Essay Answers. We defined essays to be single questions (with no subquestions) allotted 20 or more minutes, a time threshold requiring an amount of writing or with problem solving that clearly was much more than most other free-response items found in the examinations. A required essay question was found only in biology examinations from four countries (AEB, Example 3-5; UOL; Aix; Paris, Example 3-6; Israel, Example 3-7; and US) and a single chemistry examination (UOL). Essays were not required in any mathematics or physics examinations, although the Israeli physics examination included an optional one. Examinations using this item type generally had a single required essay about 20 minutes long, but some England and Wales examinations required two essays, one of which was allotted double the time—40 minutes! The occurrence of essays found in the examinations reflects the fields of study since more biological concepts readily lend themselves to long, written explanations than do the concepts of chemistry, physics and mathematics. However, it would not have been a difficult task for examination writers in the latter fields to include complex problems or questions requiring long explanations. The hurdle may, perhaps, have been the greater effort required to score essay questions.

Scientists and mathematicians generally are not known for effective written communication, a deficit that could increasingly threaten to widen the schism between them and the general population as the amount and complexity of science and mathematics knowledge explodes. Having examinations that expect college-bound students to lucidly and adequately explain phenomena or debate positions about real-world issues would tend to bridge such gaps as well as improve communication within the scientific and mathematics fields. Including essays in examinations presents issues—for example, longer testing times, expensive scoring, and unreliable scoring. But if communication is an important goal for science and mathematics instruction, it should be reflected in these high-stakes examinations.

Laboratory Practicals

Only Israel and the two examination boards in England and Wales used laboratory practicals in their examinations. (See Figures 2-1 and 2-2.) The Israeli biology practical was found to be quite different from the biology practicals in England and Wales, although they were similar in their length, administration, and percent of examination score. The Israeli practical was inquiry-oriented, requiring students to conduct an investigation. (See Example 3-8 in Chapter 3). Unlike the Israeli biology practical, however, the Israeli physics practical was not inquiry-oriented. Tamir (1995) reports that, very recently, an inquiry-oriented practical is becoming more common in the physics examinations.

The English and Welsh practicals emphasized manipulation of materials, routine procedures, and simple information. The physics practicals of the different examination boards in England and Wales had a minor difference. The AEB practicals included very precise instructions, leaving few decisions for the students to make. The UOL practicals were somewhat less prescriptive, and more explanations of actions were requested. (See sample examination questions 5-3 and 5-4 in Chapter 5).

While there is a continuing debate among policymakers and test designers on the role and importance of practical laboratory work in examinations, the literature clearly shows its importance in teaching and learning science (e.g., Tamir, 1974, 1975). Lazarowitz and Tamir (1993) document support for this position in their recent review of almost 300 studies on laboratory work in science education. Kelly and Lister (1969, p. 132) note: "Practical work involves abilities, both manual and intellectual, which are in some measure distinct from those used in non-practical work." An even stronger position is expressed by the National Science Teachers Association: "Research can show that without adequate laboratory facilities and materials most students cannot learn biology in any meaningful way" (Showalter 1986, p. 1).

In his commentary on Part I, Black identifies the scarcity of laboratory investigation in the examinations as one of the most serious concerns raised by this study. We agree and fear that an implicit message of *not* including practicals in high-stakes examinations is that one can do without them in the classroom with no serious harm done. For example, Herr (1992) surveyed over 800 Advanced Placement teachers in the United States and subsequently conducted in-depth interviews with 20 of them. He asked the teachers to compare their instruction in AP courses with their instruction in honors classes. His findings (p. 521) provide a basis for concern:

> When laboratory work is not assessed on the national AP examination, such experiences are sacrificed to provide time for lecture. When laboratory experiences are assessed, however, teachers responded by allocating more time for laboratory work.

Use of Diagrams, Photographs, Graphs and Tables

Figure 2-4 indicates that only physics examinations and, to a lesser extent, biology examinations made substantial use of non-text elements, despite the importance in all the sciences and mathematics of these alternative forms of communication and representation. About 35 and 60 percent of examination scores in biology and physics, respectively, came from items using diagrams, photographs, graphs or tables compared to only about 5 percent of chemistry and mathematics examination scores. Diagrams were twice as common as graphs and tables in biology examinations and three times more frequent in physics examinations.

Figure 2-4.   Use of Diagrams, Graphs and Tables

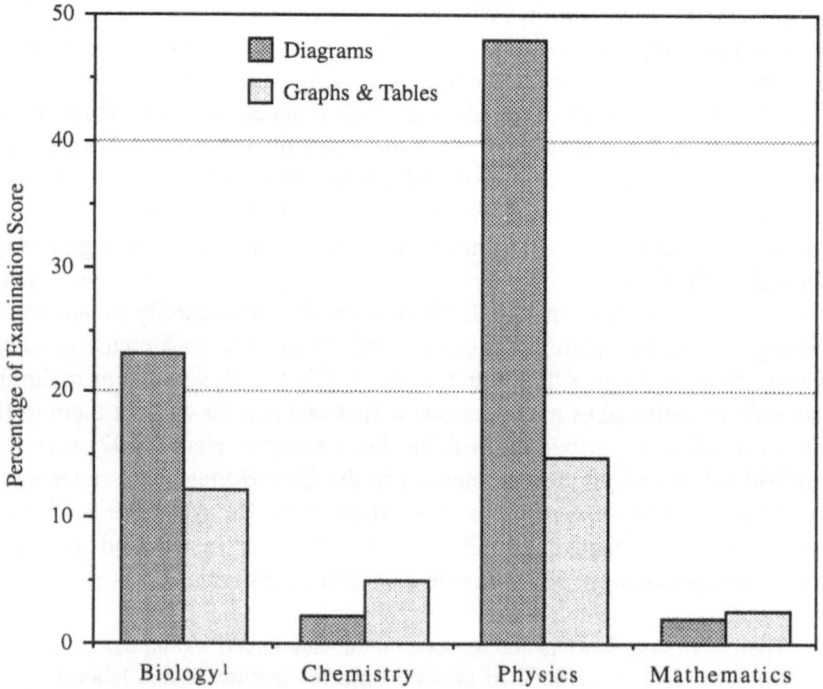

[1] Data for diagrams also include a few reproductions of photographs in the examinations of England/Wales and France.

Figure 2-4 shows aggregates; therefore it does not illustrate some interesting country variations within biology, chemistry and physics examinations that are evident from Tables 3-2, 4-2 and 5-2. (No country's mathematics examinations had notably more use of non-text elements than any other.) Among biology examinations, England/Wales and France had markedly greater use of non-text elements, including the only use of photographs, as described in Chapter 3. The heaviest use of diagrams among physics examinations was in Japan where a diagram was associated with every question. While non-text elements generally were missing from chemistry examinations, almost 20 percent of the Israeli chemistry examinations did use a graph or table.

Extensive use of non-verbal representations in examinations has substantial benefits for students and broadens the kinds of topics that can be assessed, including the nature of scientific research.[4] Examinations relying predominantly on text to convey information benefit students who can deal well with that format and penalize students who cannot. Since individual students have relatively different abilities or preferences for learning through different visuals modes as well as text, examinations that employ a variety of communication devices afford students at least some opportunity to draw upon their strengths over the course of the examination. Moreover, some topics are almost impossible to assess without diagrams, such as structures of organs and tissues in biology, or molecules in chemistry. Finally, scientists and mathematicians routinely construct or interpret graphical, tabular or diagrammatic displays as part of their work. Since scientific or mathematical investigation involves working with data in these ways, examinations with few non-text elements cannot reflect the core of what science and mathematics are all about.

Examination Topics

The examinations we analyzed generally exhibited a very conservative nature consistent with their purpose—that is, every examination primarily consisted of advanced concepts that scientists and mathematicians would expect college-bound students to know in the fields of biology, chemistry, physics, and mathematics. Chapters 3 to 6 report the examinations—topics in detail. Rather than summarize all topic coverage here, we instead note particularly interesting contrasts and omissions. For example, we point out the general absence of a fundamental topic in one of the disciplines, the

---

[4] We assume meaningful use of diagrams, not overuse of diagrams where they unnecessarily represent some very basic information that can readily and unambiguously be communicated in a few words. Among all the examinations, we only found one or two items where such unwarranted use might be judged to have happened.

omission in one country's examinations of a subdiscipline that is included everywhere else, and the relative emphases between major subdisciplines.

This discussion also points out the minimal treatment of some topics that a scientifically literate citizen might be expected to know. Admittedly, prospective college students taking these examinations probably would not need great familiarity with such topics, since college science and mathematics courses typically focus on disciplinary knowledge rather than applications to real-world phenomena.

A critical caveat to the following discussion is that it reports topics only in the two years of examinations analyzed. We do not know whether other years' examinations left out topics we found or included topics missing in the two years.

Biology

Examinations universally covered a few core topics in the field of biology, but tended to omit some others.

Topics universally emphasized	Topics generally missing
Cells	Ecology
Energy handling	Evolution
Variation and inheritance	Genetic engineering

Most biologists would argue that students should study the topics of cells, energy handling, and variation and inheritance. Additionally, however, it seems appropriate that tests should address other active areas of today's biological research, for instance, ecology, evolution, and genetic engineering. Since the mass media have been featuring stories for years about such developments as gene splicing and new human-manufactured species, it is disheartening that advanced biology examinations were not covering related science topics by 1992.

A similar argument can be made for topics in ecology and the environment. In most of the countries, an average of only 3 percent of examination questions dealt with ecology. This proportion was somewhat higher in the United States and England and Wales, where about 12 percent of examination questions addressed this subject.

Evolution typically was the focus of only 5 percent or less of examinations. Two exceptions were Germany, where 12 percent of the Bavarian examinations focused on the topic, and Israel, where evolution is one of nine biology topics from which students can choose to study six during their last two years. The minimal attention generally given evolution is surprising, since it is a core concept and an area of vigorous research in biology. Even before current standards-setting documents recommended evolution as a biology topic (American Association for the Advancement of Science, 1993; National Research Council, 1994; National Science Teachers Association, 1992), its key role had long been established by the National Research Council's Committee on High School Biology Education (1990): "A mature theory of evolution has stood up well in its broad outlines for more than a century . . . the current handling of evolution [in education] is egregious" (pp. 10, 23).

Chemistry

All chemistry examinations—except those in the United States—gave substantial attention to organic chemistry and some attention to industrial applications. No countries included environmental chemistry.

Topic	Avg. Percent of Examinations	Least Coverage
Organic/biochemistry	25%	1% - US
Industrial chemistry	8%	0% - US, SWE
Environmental chemistry	0%	0% - all

Organic chemistry is a major subdiscipline: more than half of the world's chemists devote their research to the chemistry of carbon compounds. Further, organic chemistry is just as important for a scientifically literate citizenry to understand as is inorganic chemistry—and perhaps more so, given the large and growing body of biochemical research that affects everyone's life today. There are at least two explanations for the surprising contrast between the US AP examinations and all the other examinations. Because chemistry receives more time in the secondary school curricula of other nations, both organic and inorganic topics can be included more readily. In the United States, high school chemistry typically focuses almost exclusively on inorganic chemistry. Also, the purpose of the AP examinations argues against including organic chemistry. These examinations are designed to mirror entry-level college courses, and introductory college chemistry in the United States generally is restricted to inorganic chemistry.

Since daily life increasingly is affected by industrial applications of chemistry, it is regrettable that traditional US chemistry courses only infrequently go beyond basic chemical understanding. However, *Chemistry in the Community (ChemCom)*, an increasingly popular alternative chemistry course developed by the American Chemical Society, emphasizes applied chemistry and includes some industrial processes (American Chemical Society, 1993).

Topics related to the environment are too important to the education of future citizens to be relegated to the last chapter of the chemistry book, which few classes ever get to. Despite an abundance of media stories on pressing environmental issues, such relevant topics as the threat of heavy metals in the food supply, increasing acidity and particulate matter in rain-water, negative side effects of organic herbicides and inorganic fertilizers, debates about the safety of food irradiation, and problems in disposing of radioactive waste are still lightly treated in many chemistry courses.

Physics

Some examinations were restricted solely to classical physics, while others also included modern topics.

Classical physics topics	Modern physics topics
Mechanics	Atomic and quantum physics
Electricity	Nuclear physics
Electromagnetism	Relativity
Wave Phenomena	Cosmology
Optics and light	
Thermophysics	

All examinations in England and Wales and Israel had comprehensive topic coverage in both classical and modern physics. Coverage in German, French, and US examinations varied. In France and Germany, examinations in one region restricted topics to classical concepts, while the other region's exami-nations also included modern topics. The nature of topics in AP examina-tions mirrored the content of their corresponding courses: one examination was comprehensive, while the other one was restricted to classical topics. The Japanese examinations focused almost exclusively on classical topics, as did the Swedish ones. However, the Swedish examinations are administered earlier in the year, and most classes do subsequently cover modern physics.

Even examinations with comprehensive topic coverage generally emphasize classical topics over modern ones. Since the frontiers of today's research in physics lie in the modern topics, it is surprising that some of these examinations do not emphasize modern physics more.

Mathematics

As expected, examinations focused on topics usually associated with courses known as "Precalculus" and "Calculus." Distinct patterns of topic emphasis emerged among the countries, as discussed in Chapter 6.

Universal topics	Less common or missing topics
Differentiation, integration	Probability and statistics
Functions and relations	Geometric relations
Geometric transformations	Proportionality
	Number systems
	Measurement

There was a common core of topics among the ten sets of examinations studied (five countries, with two different sets each from England and Wales, France, Germany, and the United States). All examinations focused on the topics of functions and relations and the calculus-related topics of differentiation and integration. These topics accounted for from 29 percent (Baden-Württemberg) to 100 percent (US-AB Calculus) of the examination questions. Geometric transformations, mostly viewed from a linear algebra perspective, was a third significant topic found in 7 of the 10 examinations. It constituted from 15 to 34 percent of each of the seven examination. Only four examinations included probability and statistics and geometric relations. Three topics received little to no attention in the examinations: proportionality, number systems, and measurement.

Disappointingly, all of the examinations, with the exception of special portions of the English and Welsh and German examinations, treated mathematics only as an abstract discipline. They focused heavily on symbol manipulation and only required students to recall and apply facts and definitions to solve problems similar to textbook examples. There was no evidence of modeling, attempts to connect mathematics to real-world problems, or use of other disciplines as context for mathematics. There was little evidence that any examinations involved uses of technology or types of problems reflecting calculus reform efforts which emphasize the power of computer algebra systems. These systems downplay the need for some of the manipulation-intensive integration formulae and algorithmic processes, e.g., volumes by surface of revolution techniques. Yet these problems still were prevalent in many of the examinations. From these standpoints, the examinations mostly reflected curricula unaffected by changes suggested in the *Curriculum and Evaluation Standards* published by the National Council of Teachers of Mathematics (NCTM, 1989).

Performance Expectations

In this section, we describe the operations and behaviors that the examinations required of students. Table 2-2 lists the TIMSS Student Performance Expectations that we used for this purpose. The nature of student performances in science and mathematics differs sufficiently so that separate TIMSS performance expectations have been developed for each subject. Readers already familiar with the TIMSS categories may note we have omit-

Table 2-2.   TIMSS Performance Expectations: Detailed Categories

Science	Mathematics
Understanding simple information complex information	**Knowing** representing recognizing equivalents recalling mathematical objects and properties
Theorizing, analyzing, and solving problems abstracting and deducing principles applying principles to solve quantitative problems applying principles to explain phenomenon constructing, interpreting, and applying models	**Using routine procedures** using equipment performing routine procedures using more complex procedures
Using tools, routine procedures, and scientific processes using apparatus, equipment, and computers conducting routine experimental operations gathering data organizing and representing data interpreting data	**Investigating and problem solving** formulating and clarifying problems and situations developing strategy solving predicting verifying
Investigating the Natural World designing investigations conducting investigations interpreting investigational data formulating conclusions	**Mathematical reasoning** developing notation and vocabulary developing algorithms generalizing conjecturing justifying and proving axiomatizing

ted a few performance expectations that were not applicable to these examinations. The TIMSS framework is not designed to be hierarchical, i.e., the
cognitive demands of successive categories do not necessarily increase.
Depending on the educational goals that experts advocate, however, they
often value performance categories differently. Our educational perspective
should be apparent in the following discussion of performance expectations
required by the examinations.

Expectations in Science

Table 2-3 indicates that about one-fifth to one-half of science examinations
in each country are best described by the Understanding Information category. At the extremes, England/Wales and the US especially relied on this kind
of question, while France and Sweden made little use of it. Examination
questions categorized this way require students to recall rather than process
information, e.g., students describe or compare concepts. While the examinations we analyzed certainly covered quite advanced and complex concepts,
only describing or contrasting them generally is a less sophisticated student
performance than those in other performance categories.

Table 2-3.   Performance Expectations in Science Examinations
Percentage of Examination Score

	Understanding Information	Theorizing, Analyzing, and Solving Problems	Investigating the Natural World; Using Routine Procedures
E&W	**40**	43	**24**
France	23	55	**22**
Germany	31	56	13
Israel	35	45	19
Japan	37	51	12
Sweden	22	**68**	11
US	**46**	48	5

Bold indicates highest value(s) in each column.

Swedish examinations in particular emphasized Theorizing, Analyzing, and Solving Problems (68 percent), but about half of other countries' examinations used such questions as well. Because solving quantitative problems and applying scientific principles to explain phenomena are core scientific activities, devoting a substantial portion of examinations to these performances is consistent with scientific competence.

Disappointingly, most examinations seldom required students to Investigate the Natural World or Use Routine Procedures, whether through laboratory practicals or paper-and-pencil questions. Only 5 percent of the US examinations required student performances such as gathering, organizing, or interpreting data. The English and Welsh and French examinations had the most use of these performances, 24 and 22 percent, respectively.[5] Most English and Welsh science examinations included a laboratory practical, and French examinations presented data from experiments and asked for interpretation.

Expectations in Mathematics

The data in Table 2-4 indicate that all the mathematics examinations predominantly required students to use Routine Procedures and Problem Solving, and seldom required students to perform the more sophisticated processes of Mathematical Reasoning. At least 71 and as much as 97 percent of each examination emphasized the former two performances over the latter. The Japanese examinations placed far more emphasis on reasoning (29 percent) than examinations from other countries. The two regions of France had comparable use of performances, while patterns between the two examination boards in England and Wales and the two German states differed considerably. See also Figure 6-3 in Chapter 6 that shows patterns between countries in their relative emphasis among these three general categories of performance expectations.

Seventy percent of the US AB examination expected performances in Using Routine Procedures, which generally are less demanding processes than those in Investigating and Problem Solving. Routine Procedures constituted between 35 and 52 percent of all other examinations except those from Baden-Wrttemberg that limited them to 20 percent of the examination and instead placed more emphasis on Problem Solving than other examinations (73 percent).

---

[5] If the Israeli chemistry examinations had included laboratory practicals as did the biology and physics examinations, Israel's use of this performance would have been higher than 19 percent and likely more than the French value.

Table 2-4.   Performance Expectations in Mathematics Examinations
Percentage of Examination Score

	Using Routine Procedures	Investigating, Problem Solving	Mathematical Reasoning
E&W - AEB	52	39	8
E&W - UOL	35	60	3
Fr. - Aix	38	58	3
Fr. - Paris	36	50	12
Ger. - Bav.	36	61	4
Ger. - B-W	20	**73**	6
Japan	35	36	**29**
Sweden	36	48	16
US - AB	**70**	20	8
US - BC	42	47	10

Bold indicates highest value(s) in each column.

Difficulty

Which countries' examinations are more difficult than others? Initially, the authors rated every examination item on a 3-point difficulty scale. But our consensus was that these results were unreliable. For one thing, experts felt valid difficulty ratings were impossible without considering factors surrounding the examinations, such as how students are prepared for examinations and how courses of study are linked to examinations. As Dwaine and Lucy Eubanks note in Chapter 4: "If their curricula specifically addressed the topics included in the examinations, many student responses that appeared to require advanced analytical skills may, in fact, only have required recall of specific information treated in their coursework."

Other contextual factors to be considered in appraising the difficulty of examinations include the number of examinations a student must take and pass in order to gain university admission, the procedures for administering and scoring the examinations, and more. Chapter 7 further outlines various aspects of difficulty stemming from differences in examination systems. Jack Schwille's commentary on Part II of this book also argues that an understanding of the educational and examination systems in which they are embedded is critical to comparing the difficulty of examinations.

Moreover, many internal examination features bear on comparisons of difficulty as well, and often indicate conflicting orders of difficulty. Besides sophistication of topic and complexity of treatment, these include length of the examination, degree of choice, item types used, breadth or depth of topic coverage, and student performance required. Obviously, then, a single ranking of examinations or of countries for difficulty is an unsound oversimplification. Instead, we have ranked the difficulty of countries' examinations based on some of these examination characteristics. (See Table 2-5.) In this way, a given country's examinations could be considered either more or less difficult than those of other countries, depending on the specific examination characteristic being considered.

Table 2-5.   Examination Difficulty: Features Within Examinations

For each facet of difficulty, countries are listed in decreasing order of difficulty.

How long is the exam?	E & W	6-9 hours, mostly 8
	Israel	5 to 5.5 hours
	France, Germany, Sweden	3 to 4.5 hours, mostly over 3.5
	Japan, US	2.5, 3 hours
Do students choose among questions?	Germany, Japan, Sweden	no choices
	France, US	considerable, but only in some subjects
	E & W, Israel	extensive, all subjects
What item types are used?	E & W, Israel	laboratory practicals included
	France, Germany	free-response only
	England/Wales, Israel, Japan, Sweden	free-response with some multiple-choice
	US	least free-response, most multiple-choice
How broad is the examinations' topic coverage?	E & W, Germany, Israel, US	broad
	France, Japan, Sweden	significantly fewer topics
What student performances are expected (cognitive operations)?	France, Sweden	recall only, one-fifth
	Israel, Germany	recall only, one-third
	E & W, Japan	recall only, two-fifths
	US	recall only, one-half

Curricular Implications

Every educational experience should be scrutinized as to whether it fosters learning that is consistent with educational goals. Our analysis indicates that students passing the examinations analyzed in this book certainly met the goal of learning_in considerable depth—traditional topics in the fields of biology, chemistry, physics, and mathematics. On that score, we have noted numerous interesting similarities and contrasts among the examinations.

But what about the *curricular* purposes and effects of the examinations? We found that aspects of some examinations that were analyzed are inconsistent with recommendations of curriculum specialists in science and mathematics. This is important because, as Linn (1987, p. 203) observed, teachers teach to the test and students study to the test, thus, "innovations in the curriculum fail to persist unless they are reflected in similar innovations in testing."

Potential discrepancies may arise in part because the paramount purpose of the examinations we analyzed generally is to distribute educational opportunities among prospective college students. High-stakes examinations that fulfill such needs do not necessarily meet broader curricular goals. It is possible, however, for examinations to address both curricular *and* system needs. Some examinations in this study are closer to that goal than others.

The *High Stakes of High School Science*, a book by 18 prominent educators, makes an argument prevalent in the science and mathematics education communities (National Center for Improving Science Education, 1991, pp. 33-57): Assessment should foster learning of the science and mathematics needed in the 21st century. Hence, examinations should:

- assess topics relevant to scientific or mathematical literacy,
- require a variety of student performances,
- probe higher-order reasoning, and
- assess content in the context of real-world phenomena.

Again, the examinations analyzed often were not designed to address all or even any of these US assessment goals. While we were not able to investigate systematically whether science and mathematics education communities in other nations would advocate that their examinations be evaluated against these goals, we believe that support for them is common in the countries whose examinations were included in our study. Regardless, there is considerable interest among US policymakers in learning how other nations' examinations, as well as those from the United States, fit with important US goals for curricular reform in science and mathematics.

Limited Assessment for Scientific and Mathematical Literacy

> Science, energetically pursued, can provide humanity with the knowledge of the biophysical environment and of social behavior that it needs to develop effective solutions to its global and local problems (American Association for the Advancement of Science, 1989, p. 12).

While the science and mathematics examinations discussed in this book universally assessed core topics, some gave little or no attention to scientific topics that have strong societal or global relevance (e.g., environmental science); are active areas of research in the scientific fields (e.g., genetic engineering); and represent industrial processes (e.g., large-scale production of chemicals). Examinees undoubtedly should understand core concepts at an advanced level to determine whether they have the necessary foundation for university-level science and mathematics study or careers in these fields. But not to encompass other topics necessary for scientific literacy seems unnecessarily narrow.

Because of the high stakes attached to these examinations, topics omitted from examinations are less likely to be included in science and mathematics courses. In his survey of AP teachers, Herr (1992) found differences not only in their approach to laboratory experiences, depending on what course they were teaching, but also in their general teaching strategies. Teachers reported that, compared to their teaching in honors classes without external examinations, they rushed through topics to cover everything on the AP examination, did not have time to pause and treat a topic more extensively as interest arose, and did not have the curricular freedom to delve into topics outside the syllabus.

Varied and Not-So-Varied Student Performances

> Because of the way tests are constructed, they ignore a great many kinds of knowledge and types of performance that we expect from students (Darling-Hammond, 1991, p. 220).

In addition to recalling important facts or concepts—however sophisticated they may be—students also should actively generate ideas, develop explanations of phenomena based on scientific and mathematical principles, and solve problems. Additionally, designing and conducting investigations is critical to understanding how scientific or mathematical knowledge develops, a major aim of science and mathematics education. Some countries' examinations placed much greater emphasis on a variety of performances than did others.

Little Probing of Higher-Order Reasoning

Inductive and deductive reasoning—verbal, analogical, and spatial reasoning—and creative critical thinking are among the primary elements of scientific and mathematical thinking. Yet current forms of assessment often do not stress these activities (National Center for Improving Science Education, 1991, p. 37).

Some of the examinations analyzed seldom gave students the opportunity to analyze scientific information or data and make inferences or predictions from them, evaluate or design an experiment to perform some specified function, read scientific articles, or argue points of view and support them with facts.

Few Real-World Contexts

The need for relevance in science education has always received a wide acceptance, even if its realization has proved elusive (Kesner, Hofstein, and Ben-Zvi, 1995).

To encourage students to see the connections between science and mathematics learning and real-life experiences, assessment should be placed in everyday contexts. Although such an approach is possible without compromising other testing requirements, the examinations studied usually failed to do this. Speaking of the mathematics examinations, Dossey (1994) noted that:

Reformers who subscribe to the NCTM Standards would be sorely disappointed if they looked at these exams. They are very traditional, universally placing mathematics in the abstract mode of the discipline rather than in real-world contexts, for example, drawing instead upon scientific phenomena for understanding mathematical patterns.

Summary

The data resulting from this study revealed quite large differences among the examinations from the seven countries. Such important differences in curricula have not been identified by previous international comparisons that have relied on analyses of national syllabi and/or tables of contents from textbooks. The information gleaned from syllabi and textbook introductions is mostly limited to topic information. But topic information obtained in this manner is general and often leads to the conclusion that the science and mathematics curricula of countries are more similar than different in their selection of topics. This study's more precise topic coding of examinations, made possible first by the TIMSS curricular frameworks and then by the subtopics we added to it, brought to light a pattern of diversity. The item type and performance expectations information, usually hard to discern from syllabi and textbook introductions, also indicated strong country differences.

Still, one may ask: So what? What impact do such examination differences have? Given the high stakes attached to the examinations studied, we believe that such marked differences significantly impact students' achievement, attitudes, and ways of learning. Should it cause concern that some topics treated at the college level in the U.S. are covered at the secondary level in other nations, as indicated by their examinations? If examinations predominantly restrict topics and their treatments to traditional aspects of the scientific and mathematical fields of study, do teachers have the latitude to cover topics relevant to a more general literacy needed for applying concepts to real-world phenomena? Do examinations without laboratory practicals reduce the incentive for teachers to include them in daily instruction or for students to take them seriously? If we want teachers and students to employ a variety of learning experiences, then should examinations include a variety of item types and performances rather than being overly focused on too few of them?

Similar analyses of examinations from additional countries would be useful. Further, comparing external examinations at other grade levels also could be revealing, in particular, contrasting examinations given just before or at the beginning of secondary school, such as the GCSE examinations in England/Wales, the Brevet in France, and high school entrance examinations in Japan. Analyses of these sorts of examinations could provide insight on what all students (not just the college-bound) are expected to know and be able to do at that point in their schooling.

While this chapter has compared the examinations across all subjects, the next four chapters discuss in detail the country differences among examinations within each of the four subjects of biology, chemistry, physics, and mathematics.

3

Biology
Examinations

Pinchas Tamir

Overview of Examinations

THIS ANALYSIS OF 17 BIOLOGY matriculation examinations relied on 1991 and 1992 examinations from the countries and regions under consideration (see Table 1-1), with the following exceptions:

*   Sweden has no biology matriculation examination.
*   Only the 1991 Japanese examination was included.
*   US examinations were from 1990 and 1993.

England and Wales

Advanced Level Examination, Associated Examining Board (AEB). This examination consisted of two papers and took 5.5 hours.

*   Paper 1 took 2.5 hours and had 20 questions, all compulsory. In general, this paper required mainly low-level cognitive skills, for example, matching organisms to their classes. Most questions required only a short phrase, or even one word, as an answer.

*   Paper 2 allotted 3 hours and had five extended, free-response questions, most of which contained several subquestions and all of which were compulsory. For one question, students chose from three topics to write an essay. This paper required much more sophisticated skills and analysis than did paper 1.

Advanced Level Examination, University of London (UOL) Examinations and Assessment Council. This was the longest of all examinations in the study. It had five papers and took 9 hours to complete.

- Paper 1 required 3 hours and had three sections:
  - Section A had 10 questions, mostly short-answer.
  - Section B included three extended-answer questions.
  - Section C provided 45 minutes for students to write an essay on one of two topics.

- Paper 2 took 1.5 hours and consisted of five questions, all compulsory, which focused on planning investigations, including required statistical analyses.

- Paper 3 took 2 hours. It had three sections and nine questions, all compulsory.
  - Sections A and B were a mixture of short and extended-answer items.
  - Section C allocated 45 minutes to write an essay on one of two topics.

- Paper 4 took 1 hour and provided a choice of three subjects, each having seven questions, e.g., applied animal biology, applied plant biology, or microorganisms and biotechnology.

- Paper 5 took 1.5 hours and had three extended-answer questions.

The UOL examinations had notable structural differences between 1991 and 1992. The above description generally described both years' examinations. The 1991 examination, however, had a 2.25-hour practical (laboratory) and one fewer written papers.

France

Baccalauréat Examination, Aix Region. For this 3-hour examination, students chose one of two sets of questions. However, each question set had two sections, an "Organized Recall of Knowledge" portion and an "Interpretation" portion. The former part asked students to explain a concept or solve a problem, while the questions in the other part presented data and then asked students to interpret them.

- Section 1 was an extended-answer question.

- Section 2 consisted of descriptions of experiments followed by questions about effects and conclusions.

Baccalauréat Examination, Paris Region. The examination's structures and topics were very similar to the Aix examinations. To illustrate topic choices, in one set of questions, section 1 covered cell division processes, while section 2 covered endocrine systems. In the other set of questions section 1 concentrated on the endocrine and nervous systems; section 2 examined human cell division and biochemistry. All questions on French examinations were free-response.

Germany

Abitur Examination, State of Baden-Württemberg. This examination had six parts and took 4.5 hours. The questions addressed quite sophisticated topics, went into them in considerable depth, and were free-response items. While the questions addressed advanced biology, however, the performances they required from students often were recalling facts and simple information. The six parts each had five to six questions, addressing the following topics: enzymes, immunology, genetics, neurobiology, molecular biology, and evolution.

Abitur Examination, State of Bavaria. This examination was divided into four parts and took 4 hours. The examination placed unusually strong emphasis on animal behavior, human biology, and evolution. All questions were free-response items.

- Part 1 asked students to design an experiment "to prove that a certain behavioral characteristic is inherited." Some other questions addressed cell metabolism and protein synthesis.

- Part 2 topics included mating behavior of a female spider, the mechanism of muscle contraction, introduction of new species to a certain area, bacterial reproduction, and human evolution.

- Part 3 covered learned behavior of birds, mimicry, photosynthesis, the neural mechanism of a synapse, and genetics.

- Part 4 asked students about blood sugar regulation in humans, algae growing in the North Sea, the chemical reactions of photosynthesis and respiration, and genetics.

Israel

Bagrut Examination. This 5.5-hour examination had two major components: a written (paper and pencil) component and a practical (laboratory) test. The national biology curriculum prescribed nine "basic" topics and six "in-depth" topics. Students and teachers had to choose six of the nine basic and three of the six in-depth topics, respectively. The nine basic concepts were (1) organisms in their environment, (2) metabolism of cells, (3) transportation systems, (4) communication and regulation in plants and animals, (5) reproduction with special reference to humans, (6) microorganisms, (7) evolution, (8) energy transformation, and (9) genetics. The in-depth topics change from time to time. At the time of these examinations, they were (1) animal physiology, (2) advanced genetics, (3) advanced microbiology, (4) hormonal regulation in plants, (5) animal behavior, and (6) ecology.[1]

- The written paper was comprised of four parts:
  - Section 1 offered 45 multiple-choice items, five per topic, to match the nine basic topics. Any one student responded to 30 items that covered the six topics he or she had chosen. Credit was given for 27 correct answers, allowing students to miss three items with no penalty.
  - Section 2 was free-response. Students were presented with 18 multiple-choice items (two per topic) in which the correct answer was marked. The student chose three items and explained why the marked option was the correct one.
  - Section 2 contained seven problems, of which students chose three.
  - Section 3 required students to analyze a research study that they had not seen before.

- The practical test consisted of three parts:
  - Section 1 required students to identify an unknown plant or animal with the aid of a key.
  - Section 2 was an oral examination administered and tailored to individual students based on their own long-term ecological project.[2]

---

[1] See Tamir (1985) and Doran, Lawrenz, and Helgeson (1993, pp. 401-402) for further description of the Israeli biology examination.

[2] No pertinent documents existed for this study because the oral section is specific for each student. See Tamir (1972) for further description.

- Section 3 was a 2.5-hour laboratory examination administered simultaneously to groups of 10 to 16 students. Students were presented with materials, organisms, and a problem that they had to solve by designing and performing an experiment, collecting data, making inferences, and discussing their findings.

Japan

Entrance Examination, Tokyo University. This examination had three sections, each with five questions, and took 2.5 hours. Directions such as "explain" in 100 (or 50, or 20) words required students to make concise free-response answers.

• Section 1 was a series of multiple-choice items covering an account of an experiment on photosynthesis.

• Section 2 was a mix of multiple-choice and short-answer questions about factors that influence the spread of a parasite.

• Section 3 described human development and then posed relevant questions.

United States

Advanced Placement Examination. The US examination had two papers and took 3 hours.

• Section 1 had 120 multiple-choice items, took 1.5 hours, and accounted for 60 percent of the examination's grade.

• Section 2 had four extended free-response questions, three of which were broken into subquestions. It took 1.5 hours, and accounted for 40 percent of the total grade.

General Structure

Table 3-1 summarizes key structural features of the various countries' examinations.

Table 3-1.   **General Structure of the Biology Examinations**
Number of Scorable Events (SEs)

Country/ Region	Exam Length (Hrs)	Total No. of SEs	Multiple- Choice (SEs)	Free- Response (SEs)	Laboratory Practical (SEs)
E&W - AEB	5.5	96	—	96	—
E&W - UOL	9.0	127	—	118	9
France	3.0	6	—	6	—
Germany	4.5	47	—	47	—
Israel	5.5	71	27	26	18
Japan	2.5	48	12	36	—
US	3.0	130	120	10	—

Length and Scorable Events

Length refers both to the total amount of time allocated for completing an examination and the total number of scorable events to be completed.[3] On both counts, the analyzed examinations varied greatly. The time required to complete the examinations ranged from 2.5 hours (Tokyo University) to 9 hours (UOL). The number of scorable events ranged from six (France) to 126 (US); this, of course, reflected the item types. The US examination had 120 multiple-choice items; the French examination contained only essay items.

Choice

In England and Wales, almost all the items were compulsory, but on the one or two essays students wrote, they had a choice of topics. In France, students chose between two complete examinations, each of which had the same structure and similar, although not identical, subjects. Israeli students had little choice: They selected 27 out of 30 multiple-choice questions avail-

---

[3] Scorable events were the smallest, discrete questions in each examination. They were the unit of analysis in this study and are described in the Approach to Comparing Examinations section of Chapter 1.

able on the topics they had studied. Examinations of Germany, Japan, and the United States offered students no choice on which items to answer.

Item Characteristics

This section compares item types across examinations and discusses the use of diagrams/photographs, graphs, and tables in these items.

Item Types

The amount of each examination allocated to different item types varied greatly among the set of examinations. This important examination characteristic is described three ways. Table 3-1 already provided the number of scorable events found for each main item type. Similarly, Figure 3-1 shows the amount of each main item type, but in terms of how much examination time was allotted to it.

Figure 3-1.   General Item Types in Biology Examinations

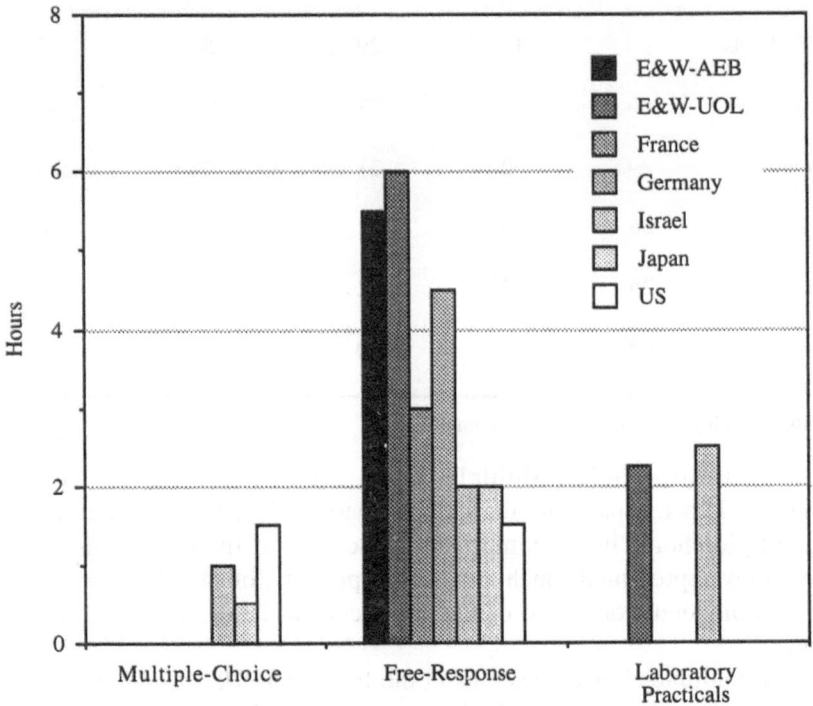

Finally, Table 3-2 shows the proportions of the total examination score accounted for by various item types. Although there were similarities—notably, that in all countries free-response items of one type or another accounted for at least a third of the total examination score—there were also major deviations. For example, the majority of items on the US examination were multiple-choice; no other country relied on this format to this extent. Similarly, one-third of the total score on the Israeli examination was derived from practical activities—again, a format little used by the other countries. If a balanced examination is one that uses a wide range of item types, each requiring different knowledge and different skills, the Israeli examination was the most balanced, and the French the least.

Table 3-2.   Item Type in Biology Examinations
Percentage of Examination Score

Country/ Region	Multiple-Choice	Free-Response, Word/ Phrase	Free-Response, Short-Answer	Free-Response, Extended-Answer	Laboratory Practical
E&W - AEB	—	13	**60**	25	—
E&W - UOL	—	12	29	52	7
France	—	—	27	**73**	—
Germany	—	2	31	**66**	—
Israel	26	—	31	7	**36**
Japan	25	**23**	12	40	—
US	**60**	—	30	10	—

Bold indicates highest value(s) in each column.

Multiple-Choice. Multiple-choice items were used only in three countries—Israel, Japan, and the United States. While in the United States, the multiple-choice items comprised 60 percent of the total examination points, they represented much smaller proportions of the other two countries' examinations: one-third of the total score in Israel, and one-fourth in Japan.

Most multiple-choice questions in the Israeli and US examinations used a common format—students had to select one of four or five answers, respectively, so the US items were slightly more difficult. The Israeli items afforded a 25 percent chance of guessing the correct answer, while the chance of guessing correctly on US items was only 20 percent. Also, for one Israeli paper, students answered 30 items, but only 27 were scored. This permitted students three errors without penalty.

The US examinations did employ a few other multiple-choice structures. Some US "matching" items, like Example 3-1, provided students with five terms, any of which could be used (once, more than once, or not at all) as answers for three questions.

The multiple-choice items in the Japanese examination differed from other examinations in two ways. First, they were dispersed throughout the examination rather than being separated in one paper or section and second, the structures of all Japanese multiple-choice items were more complex than those of Israeli and US items. For the simplest items, students had to choose one answer from six responses given instead of from four or five responses. For a matching item, students chose among eight terms instead of six. The Japanese examination contained the exceptionally complicated question seen

---

Example 3-1.   **(US, AP, 1993)**
Questions 78-80
Full Question

The group of questions below consists of five lettered headings followed by a list of numbered phrases or sentences. For each numbered phrase or sentence select the one heading that is most closely related to it. Each heading may be used once, more than once, or not at all in each group.

(A) Competitive exclusion
(B) Hardy-Weinberg equilibrium
(C) Punctuated equilibrium
(D) Allopatric speciation
(E) Character displacement

78. The allelic frequencies of a population remain constant from generation to generation.

79. This is most likely to occur as a consequence of the long-term geographic division of a population of organisms.

80. Evolution results from an event that gives rise over a relatively short time to one or more new species with traits different from those of the ancestral population, followed by periods with little or no change in the species.

---

in Example 3-2 that asked about photosynthesis in chlorophytes and photosynthetic bacteria. In two linked items, students had to choose one of three responses and then select two from eight responses. But students also had to order their three total responses in the sequence that describes the two photosynthesis processes.

Example 3-2.   **(Japan, 1991)**
Partial Question
Questions 1 and 2 precede.

Read the following descriptions, and select one correct answer from Group A and *two* correct answers from Group B. Write your answer from Group A first. List your answers from Group B in the order of the photosynthesis process, such as "1,4,5."

Group A:
1) Oxygen is discharged during photosynthesis in chlorophytes or during photosynthesis by photosynthetic bacteria.
2) Oxygen is discharged during photosynthesis in chlorophytes but not during photosynthesis by photosynthetic bacteria.
3) Oxygen is discharged during photosynthesis by photosynthetic bacteria but not during photosynthesis in chlorophytes.

Group B:
4) Oxygen, which is discharged during photosynthesis in chlorophytes or photosynthesis by photosynthetic bacteria, derives from the water.
5) Oxygen, which is discharged during photosynthesis by photosynthetic bacteria, derives from the water.
6) Oxygen, which is discharged during photosynthesis in chlorophytes, derives from the water.
7) Oxygen, which is discharged during photosynthesis in chlorophytes or photosynthesis in chlorophytes or photosynthesis by photosynthetic bacteria, derives from the carbon dioxide.
8) Oxygen, which is discharged during photosynthesis by the photosynthetic bacteria, derives from the carbon dioxide.
9) Oxygen, which is discharged during photosynthesis in chlorophytes, derives from the carbon dioxide.
10) Glucose is produced from a reaction between water and carbon in carbon dioxide.
11) Glucose is produced from a reduction of carbon dioxide by hydrogen in the water.
12) Glucose is produced through oxidation of the carbon dioxide by oxygen in the water.
13) Glucose is produced from a reaction between hydroxide ions in water and carbon dioxide.

Free-Response. The English/Welsh examinations often used short-answer items to elicit identification of particular organs, tissues, and systems presented in diagrams. These short-answer items, like multiple-choice items, made more objective scoring possible. The UOL examinations particularly relied upon short-answer items. Both the French and German examinations contained very high proportions of extended-answer items. The Japanese examination, on the other hand, appeared to avoid such answers, limiting students to "explain in 25 (or 50, 100) words," as illustrated in Example 3-3.

Example 3-3.   (Japan, 1991)
Partial Question
Questions 1-3 precede.

The weight of newborns usually decreases for 3-5 days after birth, and they regain the lost weight to the same level at about 8 to 14 days after birth. This can be explained in terms of water loss and gain. Human newborns lose water from their bodies rapidly because: (e) <u>they excrete a urine with lower specific gravity (1.002 - 1.008) than adult's urine (1.015 - 1.025), although the numbers of constituents are almost the same</u>. The keratin layer, which is the exterior most layer of skin and has a waterproof property, is still immature. Water evaporates into the air from the entire surface of the body; and a fair amount of water escapes into the air at each exhaled breath (the number of breaths in newborns: 40 - 50/minute, adults: 16 - 18/minute). On the other hand, water is gained through the intake of breast milk as well as (f) <u>water gained as the result of metabolism (that is called metabolic water)</u>.

4.   Answer regarding phrase (e). What kind of conditions are conceivable as a reason for a newborn's kidney to produce urine with a low specific gravity. Explain within 25 words by using the following as a reference:
   a.   The pituitary posterior lobe of a normal newborn has not begun to function in the same manner as adults.
   b.   If an adult has a certain illness (diabetes inspisus) due to pituitary posterior lobe failure, a large amount of urine with a specific gravity of 1.001 - 1.004 is excreted.

5.   Determine whether a - c are correct or incorrect in reference to underlined phrase (f):
   a.   Not only is a normal newborn's pituitary posterior lobe necessary for maintaining a warm body temperature, but it is also important as a source of metabolic water.

Questions 5b and 5c follow.

The free-response structure seen in Example 3-4 was unique to the Israeli biology examination. Students were provided a multiple-choice question with the correct answer identified. Their task was to explain why this answer was correct.

Example 3-4.   (Israel, 1991)
Full Question

In the next question, the correct answer is noted. Copy the answer into your notebook and explain briefly why it is correct.

2.   An accurate measurement was made between the amounts of dry substance in corn plants at two different times: at noon on a hot summer day and at midnight after that day. It is reasonable to assume that the amount of dry substance is

1. **larger at noon.**
2. larger at midnight.
3. identical at the two times.
4. in some of the plants, larger at noon, whereas larger at midnight in others.

Free-response items often were dependent upon one another. For example, item 3 asked the examination taker to explain his or her answers in item 2. Another variant was to provide certain rules and principles and ask the student to use these rules to decide whether additional statements were correct, as illustrated by question 5 in Example 3-3. One-fifth to two-thirds of examinations' free-response items were linked so that the ability to answer a question was dependent upon a correct answer to a previous question. If no allowances were made for incorrect answers to previous questions when scoring a subsequent, dependent question, such extensive use of this examination structure would increase the difficulty of examinations. Because we seldom had sufficient scoring information to determine precisely how these situations were handled, future research should look at this in detail.

Essay Items. Within the category of extended-answer items, one subset of questions—essay items—was substantially longer than the others. Essays were defined here as single questions that take at least 20 minutes to answer. Among the analyzed biology examinations, those from Germany and Japan did not have any essay questions.

•   The AEB examination asked students to write a 40-minute essay like the one in Example 3-5, where students chose from among three questions.

The UOL examination required students to produce two 40-minute essays, with a choice of two topics for each essay.

- In France, a single essay question comprised half of the examination score. Students chose between two questions like the one in Example 3-6.

- The Israeli examinations included an essay like the one in Example 3-7.

- The US examinations featured one 25-minute essay question in their free-response sections that comprised 40 percent of the examination score.

Example 3-5.   (England and Wales, Associated Examining Board, 1992)
Full Question

Write an essay on one of the three topics listed below. Credit will be given not only for biological accuracy but also for the organization and presentation of the essay.

A.   Locomotory adaptations to aquatic and aerial life
B.   Gametes and their formation
C.   The use of gene manipulation to produce recombinant DNA

Example 3-6   (France, Paris region, 1991)
Full question

Show that the hypothalamus is the integration center in fighting cold. In doing this, explain the process of integration of afferent messages at the level of a neuron of this center, and using the example of an effector controlled by hormones, show that the hypothalamus participates in maintaining body temperature in response to cold by adapting the response of the effector selected.

Example 3-7   (Israel, 1992)
Full question

It is argued that marrying relatives is not desirable. At the same time, it is known that isolated communities of certain species continue to survive for many generations. How can this apparent contradiction be accounted for?

Laboratory Practical. The laboratory practicals in the two examinations that used them—UOL and Israel—had much in common: examinees were allotted similar time (135 and 140 minutes, respectively); students worked individually and answered all questions in writing so that the assessment could be done in a central place at a scheduled time; they accounted for 20 and 30 percent of the final examination score, respectively. However, there was a fundamental difference between the two. The A-level practical emphasized the manipulation of equipment, routine procedures, and simple information.[4] The Israeli practicals were inquiry oriented, requiring the student to perform a whole investigation. They demanded higher-level skills in a scientific, problem-solving context. (See Example 3-8.)

Example 3-8   (Israel, 1991)
Full question

Part A.
The substance methylene blue loses its blue color in a reduced environment. Such an environment is formed when oxidization processes take place. In this part of the test, you will examine the reaction of methylene blue in different biological environments.

1. a. Mark two test tubes with the letters a, b.
    b. Mix the yeast suspension on your desk and pour 3 ml of it into each of the test tubes.
    c. Prepare the water bath at a temperature of 500 C (range: 450 - 550C). There is hot and cold water at your disposal.
    d. Add 0.5 ml water to test tube a. Add 0.5 ml detergent solution (1% active substance) to test tube b.
    e. Mix the contents of the test tubes, put them into the water bath and wait for about one minute in order to equalize the temperatures.
    f. Add 2 drops of methylene blue solution into each test tube and mix by shaking. If the color in one of the test tubes disappears immediately, add one more drop to the two test tubes. Leave test tubes in the bath, and write down the time.

2. a. Observe the test tubes every 2 minutes until you do not detect any color changes any more (but no more than 10 minutes).
    b. Sum up your observations.

3. Suggest an explanation for the results of your observations. Plan an experiment to test your suggested explanation, or a certain aspect of that explanation. In your planning you are not limited by either methods, substances or tools. Describe the experimental design by answering the following questions:

4. What is the hypothesis to be tested in the experiment?
5. What is the dependent variable?
6. How will you measure the dependent variable?

---

[4] The 1991 University of London laboratory practical in biology is reprinted in Gandal et al., 1994, pp. 25-28.

Example 3-8   Continued

7.   What is the independent variable?
8.   How will you change the independent variable?
9.   List additional important experimental elements which are related to the design you have suggested.

Hand Part A to the examiner and obtain Part B, with the necessary equipment.

Part B
In this part of the test you will examine the effect of different concentrations of detergent on yeast respiration. Before you start working, read until the end of Item 11 and prepare the table required in Item 12.

10. a.   On your desk you will find distilled water and a 1% detergent solution. Prepare 10 ml of the detergent solution where the concentration of active substance is 0.25%.
    b.   Describe how you prepared the solution.

11. a.   Prepare a water bath at a temperature of 500C (range 450-550). There is hot and cold water at your disposal.
    b.   Prepare a set of 5 test tubes according to the instructions below:

Number of Test Tube	Yeast Suspension	Detergent 0.25%	Water
1	3 ml	—	2 ml
2	3 ml	0.5 ml	1.5 ml
3	3 ml	1 ml	1 ml
4	3 ml	2 ml	—
5	3 ml	—	2 ml

    c.   Put the 5 test tubes into the bath and wait for about one minute to equalize the temperatures.
    d.   Add 2 drops of methylene blue to test tubes 1-4 (but not to test tube 5!) and mix. If the color in one of the test tubes disappears immediately, add another drop of methylene blue to each of the four test tubes.
    e.   Write down the time.
    f.   Observe the test tubes every 2 minutes until you do not detect any more color changes (but no more than 20 minutes). If a blue ring or a blue deposit appear, ignore them.

12. Write down your observations in the table. For each test tube, state how much time has passed until no more color change took place. If the time duration is more than 20 minutes, write down >20.

13. The summary of the results of the experiment can be presented either by a line graph or by a bar diagram.
    a.   Which will you prefer? Explain.
    b.   Draw the results in the way you have chosen. There is millimetric paper at your disposal.

In addition to requiring laboratory activities, the English and Welsh examinations also used "dry lab" questions to assess understanding of experimental methods, i.e., they presented a scientific problem, asked students to design an experiment that would investigate it, but did not require students to actually conduct the experiment. (See Example 3-9.)

Example 3-9   (England and Wales, Associated Examining Board, 1991)
Full question

Beetroot cells contain a red, water-soluble pigment. Detergents are said to affect the permeability of cell membranes. You are required to design an experiment to find out which of two liquid detergents has the greatest effect on the permeability of beetroot cell membranes.

a. Describe how you would set up the experiment.
b. Give three features that it would be necessary to keep constant. In each case, describe how you would do this.
c. How would you assess the effect of the detergents on the beetroot?
d. Draw a table with appropriate headings suitable for recording the raw results of the investigation.

Use of Diagrams/Photos, Graphs, and Tables

As shown in Figure 3-2, the examinations of England and Wales and France incorporated more graphics (photographs, diagrams, graphs, and tables) than other countries' examinations. Both countries occasionally used reproductions of photographs to construct questions, usually once per examination. For example, photographic images of chromosomes were found in both countries' examinations. Both AEB and UOL examinations reproduced electronmicrographs and asked students to identify structures—for example, a frog's retina or muscle tissue—and explain their appearances.

While the frequency of graphics was comparable for England and Wales and France, the variety and extent of these devices in the England and Wales examinations was particularly impressive. Three times longer than French examinations, the England and Wales examinations were saturated throughout with graphs and tables of data, diagrams of animal structure, etc. Just one of five papers in a UOL examination contained 13 pictures or drawings of organisms or tissues, nine tables and nine graphs of data or relationships between variables, and three drawings of laboratory apparatus. It would not have been surprising if such long examinations had placed greater reliance on text only, given the cost of using graphics such as creating them and having to reproduce and administer an examination with more pages.

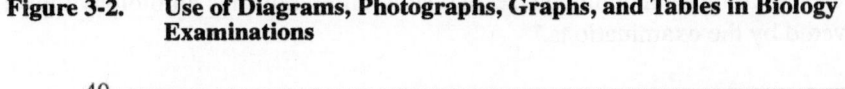

Figure 3-2.   Use of Diagrams, Photographs, Graphs, and Tables in Biology Examinations

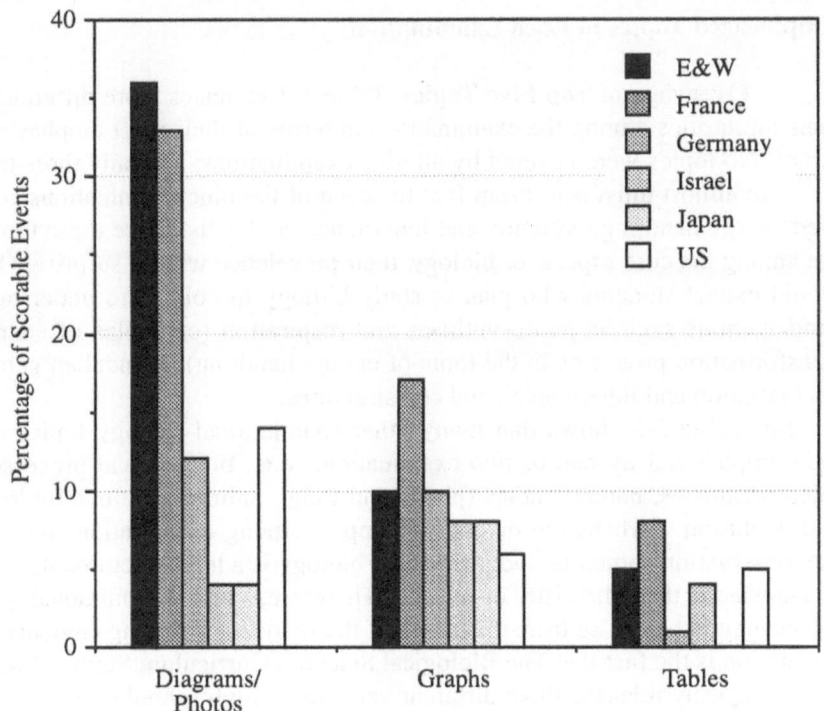

The nature of biology and student learning suggest a need for graphics. The richness of biology is difficult to express in words alone. For example, examinations with graphics can better assess students' understanding of some important biological topics such as structures of organs and tissues. Since using data in tables and graphs is a mainstay of scientific research, it is important to assess student ability in these activities. As for student learning, graphics can help students who have problems in reading.

Examination Topics

The following discussion compares the examinations' topics in two ways. The first subsection notes contrasts only among the most emphasized topics of each examination, while the second subsection indicates all biology topics covered by the examinations.[5]

Emphasized Topics in Each Examination

Overview of Top Five Topics. Table 3-3 indicates more differences than similarities among the examinations in terms of their most emphasized topics. No topics were covered by all nine examinations, and only three topics were almost universal. From five to seven of the nine examinations covered energy handling, variation and inheritance, and cells. Since these topics are among the core aspects of biology, their prevalence was no surprise. One would expect students who plan to study biology in college to understand fundamentals such as photosynthesis and respiration (examples of energy transformation processes in the topic of energy handling), Mendelian genetics (variation and inheritance), and cell structures.

But Table 3-3 shows that many other foundational biology topics are only emphasized by one or two examinations, e.g., biochemical processes, microorganisms, nature studies (plants and fungi, animals), human biology, and evolution. Perhaps the diversity of topics among examinations reflects the observation sometimes advanced that biology is a less structured domain of knowledge than chemistry or physics. Therefore, a lack of commonality in topics may in part arise from the nature of the field. Also lending credence to this notion is the fact that The Biological Sciences Curriculum Study (BSCS) simultaneously released three different versions of high school biology textbooks when it revamped biology education in the sixties.

The emphasis in German (Bavarian) examinations on animal behavior was surprising but understandable. The subdiscipline of ethology, illustrated by Example 3-10, generally does not receive a lot attention in the field of biology. Since some of the prominent researchers in ethology are German,[6] however, placing emphasis on this topic is consistent with national perspectives.

---

[5] Appendix A may be helpful for understanding the meaning of the topics because it lists subtopics associated with them. Topics were taken from the Science Framework of the Third International Mathematics and Science Study (Robitaille et al., 1993).

[6] For example, Konrad Lorentz studied imprinting behavior in geese.

Table 3-3.   Biology Topics: Five Most Emphasized in Each Examination
Percentage of examination score

	AEB	UOL	Aix	Paris	Bav	BW	Isr	Jap	US	Topic Frequency*
Energy Handling	11	12	41		23		19	29		6
Variation & inheritance		10		24	19	22	11	9	11	7
Cells	10	14		20		15	17			5
Sensing & responding			27	14	8		11			4
Biochemical processes	14		8	35						3
Reproduction				8			14	15	10	4
Organs & tissues	11							7	10	3
Microorganisms		8				12				2
Interdependence	11								11	2
Plants & fungi		8							13	2
Animals						15				1
Life cycles				8	7					2
Human biology						10		13		2
Evolution, speciation					12					1
Animal behavior					9					1
Total—Topic Concentration**	57	52	92	100	71	74	72	73	55	

Blank cells only indicate that the topic was not one of the top five in the examination. The examination still may have included the topic.

*   Topic Frequency indicates the number of examinations having the topic.

**   Topic Concentration indicates the percentage of the examination comprised by its five most emphasized topics.

Example 3-10.   (Germany, Bavaria, 1991)
                   Full Question

2.   An ant-eating beetle waits beneath a funnel-shaped sand hollow for its prey.
      Should an ant crawl into it, loose sand slides out from under the ants' legs as
      the ant tries to escape. In addition, the ant is pelted with sand by the beetle,
      usually causing it to lose its footing and tumble down. The beetle clasps its
      prey with its sucking pinchers and sucks it dry of nutrients.

Using relevant technical terms from the field of ethology, explain the behavior of
the beetle!

Topic Concentration. Interesting differences arose in topic concentration—the percentage of the examination comprised by its five most emphasized topics.[7] The German, Israeli, and Japanese examinations had topic concentrations of between 71 and 74 percent, while emphasized topics only represented 52 to 57 percent of examinations in England and Wales and the United States. Conversely, at least 43 percent of the US and English and Welsh examinations were available for other, less emphasized topics. Only 26 percent or less of the German, Israeli, or Japanese examinations addressed topics outside their emphases. While the US and English and Welsh examinations had the greatest number of questions, they still could have focused on a few topics; instead, they used their numerous questions to cover a larger number of topics. The fact that emphasized topics constituted 92 to 100 percent of French examinations resulted from their general structure. Since French examinations consisted entirely of free-response items requiring lengthy responses, they could only cover a few topics.

All Topics Covered by Each Examination

Most Common Topics. Like Table 3-3, the data in Table 3-4 show that most examinations covered the following five core topics in biology (each followed by its average percentage of examination coverage): energy handling, 22; variation and inheritance, 18; cells, 14; sensing and responding, 13; and biochemical process, 12. No other topics approached this amount of treatment: The next topics, in order of emphasis, only averaged 5 to 7 percent of the examinations.

---

[7] This statistic can be thought of as an indicator of an examination's breadth in topic coverage. Examinations with higher and lower topic concentrations focused on fewer or more topics, respectively.

Table 3-4. **Biology Topics: All Topics Found in Each Examination**
Percentage of Examination Score

	AEB	UOL	Aix	Paris	Bav	B-W	Israel	Japan	US
Diversity, organization, structure of living things									
Plants, fungi	6-10	—	—	—	1-5	6-10	1-5	1-5	6-10
Animals	6-10	6-10	—	—	6-10	24	1-5	6-10	1-5
Other organisms	1-5	14	—	—	1-5	19	1-5	—	1-5
Organs, tissues	14	6-10	—	—	1-5	6-10	6-10	6-10	6-10
Cells	15	17	—	—	6-10	25	16	6-10	14
Life processes and systems enabling life function									
Energy handling	13	20	51	—	27	6-10	21	42	13
Sensing and responding	11	1-5	35	20	11	6-10	11	6-10	6-10
Biochemical processes in cells	20	6-10	6-10	48	1-5	1-5	6-10	1-5	1-5
Life spirals, genetic continuity									
Life cycles	1-5	6-10	6-10	6-10	—	1-5	1-5	1-5	6-10
Reproduction	1-5	6-10	6-10	_	1-5	1-5	12	21	6-10
Variation and inheritance	6-10	6-10	6-10	33	24	36	13	13	15
Evolution, speciation, diversity	1-5	1-5	1-5	—	15	1-5	1-5	—	1-5
Biochemistry of genetics	—	1-5	—	—	—	—	1-5	—	1-5

Table 3-4.   Continued

	AEB	UOL	Aix	Paris	Bav	B-W	Israel	Japan	US
Interactions of living things									
Biomes and ecosystems	—	—	—	—	—	—	—	6-10	—
Habitats and niches	1-5	—	—	—	—	—	—	—	1-5
Interdependence of life	**14**	6-10	—	—	1-5	—	1-5	—	6-10
Animal behavior	—	—	—	—	6-10	1-5	—	—	1-5
Human biology and health									
	18	6-10	—	—	**16**	—	1-5	—	6-10

Bold indicates values greater than 10 percent. Columns may exceed 100 percent because some scorable events cover more than one topic. Table values 1-5 and 6-10 indicate that average values for two years of these examinations fell within these ranges. Averages less than 0.5 percent are not included.

From the most general topic level of Table 3-4, it is apparent that more examinations covered the two topics of Life Processes and Life Spirals and Genetic Continuity more than other topics, and these topics constituted the greatest percentages of the examinations. The attention to Diversity, Organization, and Structure of Living Things (nature studies) was not as strong, even though this is a traditional part of biological studies.

Less Common Topics. Few examinations covered Interactions of Living Things (ecology) even though understanding such interactions in the environment is increasingly critical to the quality of life. The US and English and Welsh examinations placed the most emphasis on ecology, and Examples 3-11 and 3-12 illustrate their treatment of it.

The amount of attention to Human Biology and Health (including disease and nutrition) was uneven, with the AEB and Bavarian examinations allocating 18 and 16 percent of the examination score to it. This area of biology most obviously is relevant to students' lives. But, at least in the United States, human biology has a vague place in the secondary school curriculum that could explain its weak presence in the examinations. Sometimes biology

Example 3-11.   (England/Wales, Associated Examining Board, 1991)
Full Question

14. Insectivorous plants such as the pitcher plant are often found in areas where
the soil is poor in nutrients. Insects are lured into the pitcher where they slip
on the smooth waxy lip and drown in the pool of liquid in the bottom.
Glands inside the pitcher secrete digestive enzymes which slowly digest the
dead insects. Some organisms are resistant to the enzymes and are able to
live in the pitcher liquid. The diagram shows a pitcher and a simplified food
web showing some of the organisms living in it.

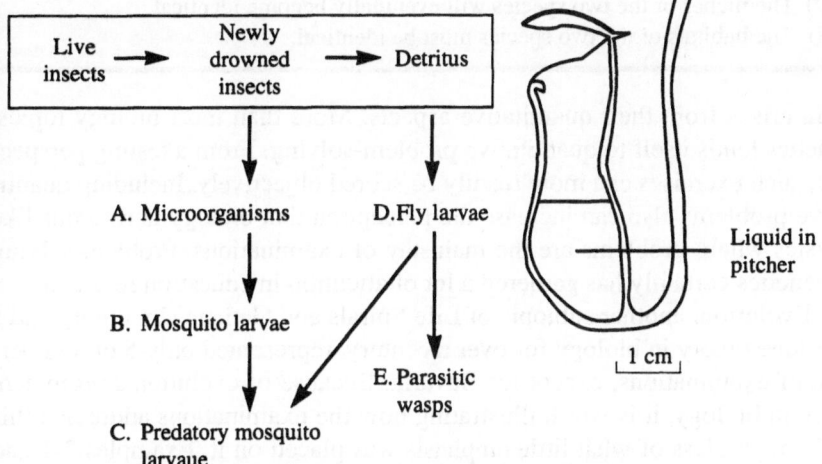

(a) Give one possible advantage to the pitcher plant of catching and digesting
insects.

(b) With reference to the information given in this question, explain what is
meant by a community.

(c) In the space below, draw a labelled diagram to represent a pyramid of numbers for the food chain A Æ B Æ C

(d) Suggest an explanation for the fact that, at any one time in this food web,
the biomass of microrganisms (A) is less than the combined biomass of the
fly larvae (D) and mosquito larvae (B).

classes include it; in other cases, separate health courses cover it. We do not
know if this situation exists in other countries.

While the more general topic of Life Spirals and Genetic Continuity was
strongly represented in these examinations, most attention was placed on the
specific topics of variations and inheritance (genetics) or reproduction. These
are important biological concepts, but perhaps the heightened attention to

Example 3-12.   (US, AP, 1993)
Full Question

32. If two closely related species exist in the same ecosystem, which of the following statements is most likely true?

(A) The more the niches of the two species overlap, the more likely it is that one of the two species will be eliminated from the area.
(B) If one of the species becomes extinct in the area, the other will probably also become extinct in the area.
(C) The two species will eventually hybridize and become a single species.
(D) The niches of the two species will eventually become identical.
(E) The habitats of the two species must be identical.

them arises from their quantitative aspects. More than most biology topics, genetics lends itself to quantitative problem-solving. From a testing perspective, such exercises can more readily be scored objectively. Including quantitative problems also can increase the perception that biology is rigorous like physics where problems are the mainstay of examinations. Problem-solving in genetics certainly has garnered a lot of attention in education research.

Evolution, another subtopic of Life Spirals and Genetic Continuity and a keystone theory in biology for over a century, represented only 5 or less percent of examinations, except for Bavaria. Because of evolution's prominent place in biology, it is worth illustrating how the examinations addressed this topic, regardless of what little emphasis was placed on it. Examples 3-1 and 3-7 came from US and Israeli examinations, and Example 3-13 is from the AEB. German examinations placed the greatest emphasis on evolution: Examples 3-14 and 3-15 are from Bavaria and Baden-Württemberg.

Another specific topic of Life Spirals and Genetic Continuity receiving little attention was Biochemistry of Genetics (genetic engineering). It seems inappropriate that by 1991-1993, these examinations failed to give much attention to such a burgeoning area of biological research. This research already is having marked impacts on daily life, and its influence will only grow. Examinations in England and Wales and the US did touch upon the topic, as illustrated by Examples 3-5 (part c), and 3-16 (part c), respectively.

Example 3-13.   **(England/Wales, Associated Examining Board, 1991)**
                    Full Question

18. The graphs show three basic types of natural selection. The shaded areas
    marked with arrows show the individuals in the populations which are being
    selected against.

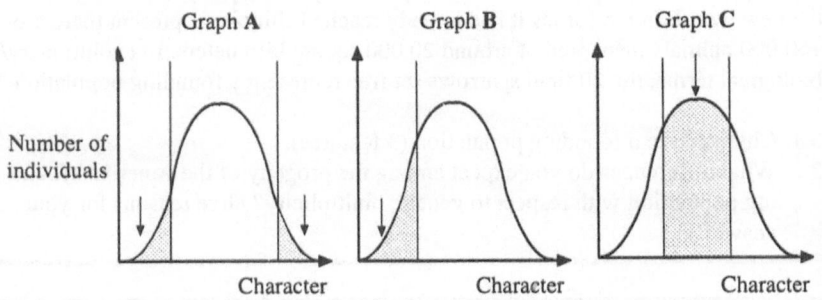

Graph **A**   Graph **B**   Graph **C**

Number of
individuals

Character   Character   Character

(a) What name is given to the type of selection shown in graph A?
(b) Describe one specific example of the type of selection shown in graph B.
     In your answer:
         (i) name the organism:
         (ii) describe the character selected.
(c) What will happen to the modal class in subsequent generations as a result of
     the type of selection shown in:
         graph B:
         graph C?

Example 3-14.   **(Germany, Bavaria, 1991)**
                    Full Question

8. When the teacher John Scopes, of Dayton, Ohio, tried in 1925 to teach his
    students about the theory of evolution and the animal origins of humans, he
    was forced to stand trial and was ultimately convicted.

    Give examples from two relevant fields of research which support the theory
    that humans and primates are related.

Example 3-15.   (Germany, Baden-Württemberg, 1992)
Partial Question
Question 2 precedes.

On April 24, 1870, in Lafayette Park, St. Louis, Missouri, a bird seller set free 20 field sparrows which had come from Germany. At first the tiny population grew in and around St. Louis. Then the field sparrow spread rapidly northwards into the new world. In the forties it had already reached Illinois; at present there are 150,000 animals in an area of around 20,000 square kilometers. In evolutionary/ biological terms, the 20 field sparrows set free represent a founding population.

3.1 Characterize a founding population (3 features).
3.2 What differences do you expect among the progeny of the American found- ing population with respect to genetic multiplicity? Give reasons for your answers.

Example 3-16.   (US, 1993)
Full Question

4.   Assume that a particular genetic condition in a mammalian species causes an inability to digest starch. This disorder occurs with equal frequency in males and females. In most cases, neither parent of affected offspring has the condition.

(a) Describe the most probable pattern of inheritance for this condition. Explain your reasoning. Include in your discussion a sample cross(es) sufficient to verify your proposed pattern.
(b) Explain how mutation could cause this inability to digest starch.
(c) Describe how modern techniques of molecular biology could be used to determine whether the mutant allele is present in a given individual.

Scale and Time Dimensions.[8] For two final perspectives on topic coverage, consider the size (or scale) and time dimensions of biological phe- nomena. The following list of examinations is ordered by the size of phenomena they emphasized, beginning with those that focused on the smallest scale (molecules) and progressing to examinations that emphasized larger scales (ecosystems). The French examinations focused on the most micro aspects of biology, while the English and Welsh and US examinations emphasized macro aspects. To illustrate the French examinations' degree and nature of emphasis on micro phenomena, half of the 1992 Paris examination

[8] These perspectives and comments were offered by Barbara Buckley, Boston, MA.

presented a study of the effects of two hormones on target cells and asked students to interpret the data.[9] More specifically, the research presented in the examination described the effects of insulin on fat cells and that of a gastric hormone, bombazine, on the pancreatic acinar cells that secrete pancreatic juices.

<u>Size of biological phenomena most emphasized</u>

France	molecular, subcellular, and cellular
Israel, Japan	subcellular, cellular, suborganism
Germany	cells, organism
E&W, US	organisms, ecosystems

The time dimension ranged from a focus on longer-term processes like evolution and inheritance to shorter, dynamic processes such as reproduction or energy handling. Most examinations had more coverage of the shorter-term processes than of the longer-term phenomena.

The leading edges of biological research lie at the ends of the scales—molecular biology and genetics, ecological studies, and evolution studied within the context of environmental changes. Considerable research also focuses on disease at the molecular level. For students to be well-prepared for future study in biology, they should understand both ends and examinations should cover them. Most examinations had some attention to both, particularly the examinations of England and Wales, but the French examinations concentrated on microprocesses to the exclusion of macrophenomena.

Performance Expectations

Figure 3-3 illustrates the general categories of performance expectations found in the various biology examinations.[10] The most prominent observation is that Understanding Information questions dominated all the biology examinations except those from France. Such questions merely require students to recall information, however sophisticated the concepts involved may be. Overall, examinations placed some emphasis on Investigating the Natural World, gave limited attention to Theorizing, Analyzing and Solving Problems, and essentially ignored Using Routine Procedures, and Science Processes. The French and English/Welsh examinations, however, gave con-

---

[9] This examination is printed in *What College-Bound Students Are Expected to Know About Biology* (Gandal et al., 1994).

[10] Performance Expectations, categories used in the Third International Mathematics and Science Study to describe the student behaviors elicited by various types of questions, are further described in the text accompanying Tables 1-3 and 2-2.

Figure 3-3.   General Performance Expectations in Biology Examinations

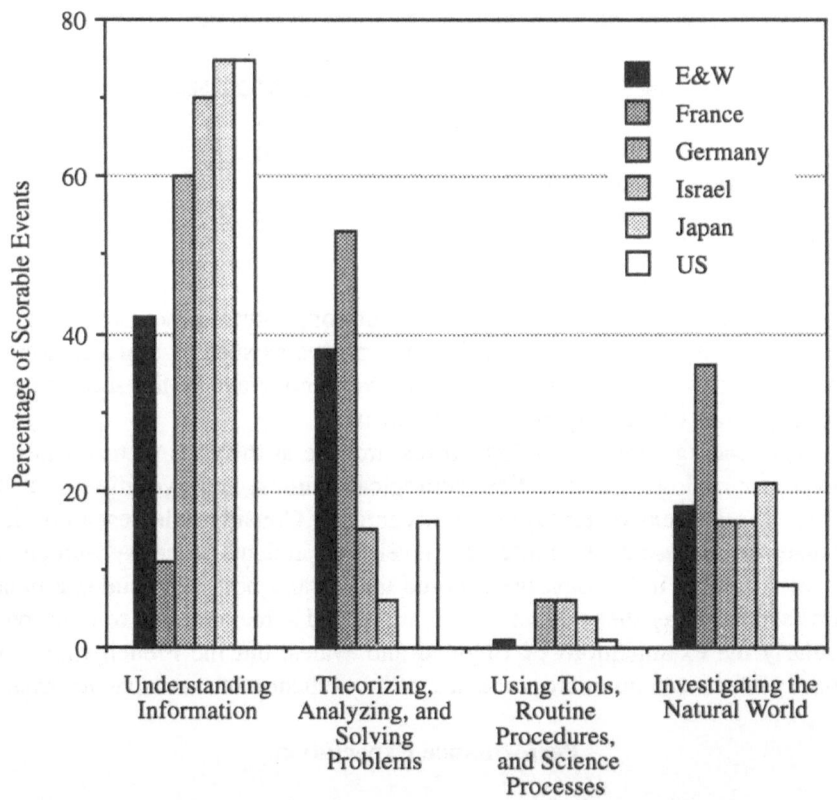

siderably more attention to questions in the Theorizing and Analyzing category while Japanese examinations did not contain any questions addressing this category.

Table 3-5 describes the performance expectations in greater detail. Regarding the Understanding category, the relative amounts of simple versus complex information addressed by the examinations differed by country. The US examinations by far had the greatest reliance on simple information (46 percent), followed by examinations from England and Wales (27 percent). On average, only 10 percent of the other countries' examinations were limited to simple information. Conversely, the emphasis that German, Israeli, and Japanese examinations placed on more complex information (54-57 percent) was several times greater than the attention given to this category by other examinations.

Table 3-5.   **Detailed Performance Expectations in Biology Examinations**
Percentage of Examination Score

Expectation Category	E&W	France	Germany	Israel	Japan	US
Understanding						
Simple Information	27	—	5	13	19	46
Complex Information	14	7	54	57	56	28
Theorizing, Analyzing, and Solving Problems						
Abstraction/ Deduction	—	17	—	—	—	—
Solving Quantitative	2	1	4	2	—	4
Developing Explanations	36	20	11	5	—	8
Constructing, Using Models	—	15	—	—	—	4
Using Tools, Routine Procedures, and Science Processes						
Apparatus, Equipment	—	—	—	—	—	—
Routine Experiments	—	—	—	—	—	—
Gathering Data	—	—	—	—	—	—
Organizing, Representing Data	—	—	1	—	—	—
Interpreting Data	—	—	5	6	4	—
Investigating the Natural World						
Identifying Questions	—	—	—	2	—	—
Designing Investigations	3	1	2	2	—	1
Conducting Investigations	—	—	—	—	—	3
Interpreting Data	12	35	8	2	8	4
Formulating Conclusions	3	0	5	10	13	—

As for details of the Theorizing, Analyzing and Solving Problems category, only French examinations required the sophisticated thinking required to abstract and deduce concepts from data. Similarly, only French and US examinations expected students to construct models based on information or data provided. In contrast, all examinations except the one from Japan expected the somewhat less but still challenging performances of developing explanations for biological phenomena presented in problems. It was not a surprise to find that biology examinations put little emphasis on solving quantitative problems.

Regarding the category of Investigating the Natural World, most examinations had questions for designing investigations, interpreting data, or developing conclusions. These performances were inherent to the laboratory practical in England/Wales and Israel, but other countries also addressed them, although only through experimental situations presented on paper. As research indicates, assessing the ability to design and carry out scientific investigations through these two very different means (hands-on and paper-and-pencil) does not necessarily tap the same student knowledge and skills. The independent research portion of the Israeli examinations was the only place where students were expected to identify a research question for investigation.

Summary

The data resulting from this study revealed tremendous differences among the biology examinations from six countries. The exceptionally long examinations of England/Wales had a greater focus on nature studies than other examinations, allotting considerable time to questions on animals, organs and tissues. But they also emphasized biochemical processes. Questions tended to address simple rather than complex information but employed many, varied graphical devices to do so. The French examinations stressed micro processes of energy handling as well as biochemical processes in cells. The level of information was more complex than simple, and a considerable amount of non-text elements were used. German examinations had more diverse topic coverage than the French examinations and were the only examinations with an amount of emphasis on evolution commensurate with its biological importance.

The Israeli examinations had broad topic coverage and gave similar attention to both simple and complex information. They put a stronger emphasis on laboratory investigation than did the University of London examination (1991), the only other examination to include this item type. The Japanese examination also covered a wide variety of topics but paid special attention to energy handling. It emphasized complex information, but was the only examination to omit various aspects of problem solving. The US examinations had the greatest emphasis of any examination on simple information. While they offered the broadest selection of topics, heavy use of multiple-choice items limited the depth to which many of them could be assessed.

We have provided here some of the more notable generalizations about each country's examinations, despite some concern about the hazard of misrepresenting a country's examinations by omitting their exceptions to the rule. We urge readers to be judicious in handling these summaries and to refer to the chapter for detail.

4

Chemistry Examinations

Dwaine Eubanks
Lucy Eubanks

Overview of Examinations

THIS ANALYSIS OF CHEMISTRY matriculation examinations relied on 1991 and 1992 examinations from the countries and regions under consideration (see Table 1-1), with the following exceptions:

- Only the 1991 Israeli examination was included, as no translation of the 1992 examination was available.

- United States examinations were from 1989 and 1993.

England and Wales (1991, 1992)

The English and Welsh examinations were at one extreme among the set of examinations in this study. They were quite comprehensive, incorporated a variety of item types, and expected students to demonstrate practical skills as well as to manipulate facts and theories of chemistry.

Advanced Level Examination, Associated Examination Board (AEB). This examination had four papers and allotted a total of 8.25 hours.

- Paper 1 allotted 2.25 hours and consisted of nine free-response questions which were broken into several subquestions. Students answered the first question and three of the remaining eight questions.

- Paper 2 allotted 1.75 hours, with seven free-response questions broken into several subquestions.

- Paper 3 allotted 1.25 hours and had 40 multiple-choice questions.

- Paper 4 allotted 3 hours and consisted of three laboratory tasks for students to complete by following specific instructions. These tasks were a titration series and tests to identify an inorganic and an organic unknown.

Advanced Level Examination, University of London (UOL) Examinations and Assessment Council. This examination had four papers and was allotted 8.5 hours.

- Paper 1 allotted 2 hours and consisted of seven free-response questions broken into several subquestions.

- Paper 2 allotted 2.5 hours and had four free-response questions consisting of several subquestions. Students answered all of the first question, one of two options in the second question, one of four options in the third question, and one of two options in the fourth question. All the questions were free-response, and each option for question 3 required an essay which was allotted 35 to 40 minutes.

- Paper 3 allotted 1 hour and had 40 multiple-choice questions.

- Paper 4 allotted 3 hours and consisted of three laboratory tasks similar to those in the AEB examinations. Some examination boards included a practical component completed during the course and assessed by the student's teacher.

France (1991, 1992)

The French examinations were at the other extreme to the English and Welsh examinations. French students were tested in some depth on a very limited selection of topics, which were presumably drawn from a much broader universe of curriculum objectives for students at that level. The underlying philosophy here appeared to be that a student's performance on the narrow range of topics would have accurately reflected the performance of that student on a comprehensive examination.The two French examinations were quite similar. In both, chemistry and physics items were presented as a combined examination which allotted 3.5 hours; the chemistry portion allotted nearly 1.5 hours. On each examination, there were two chemistry sections and three physics sections.

Baccalauréat Examination, Aix Region. In 1992, one section dealt with conjugate acid/base pairs; the other covered oxidation of an alcohol, hydration of alkenes, polymerization, and esterification. Both sections had four questions broken into subquestions.

Baccalauréat Examination, Paris Region. In 1992, one section dealt with buffer solutions; the other dealt with the effect of temperature on the decomposition of hydrogen peroxide. Each section had two questions consisting of several subquestions.

Germany (1991, 1992)

Abitur Examination, State of Baden-Württemberg. This examination allotted 4 hours. It had three sections, each devoted to a group of topics, and was composed of two free-response questions broken into subquestions.

Abitur Examination, State of Bavaria. This examination allotted 4 hours. It had four sections, each devoted to a group of topics, and was composed of four free-response questions broken into subquestions.

Israel (1991)

Bagrut Examination. This examination had two papers and allotted a total of 5 hours.

• Paper 1, allotted 2.5 hours, was compulsory for both low-level and advanced-level chemistry students. It was the only paper required for the low-level students. This paper had 16 multiple-choice questions and six free-response questions, each broken into subquestions. Students chose to answer any three of the free-response questions.

• Paper 2, allotted 2.5 hours, was required for only the advanced-level chemistry students. Students first answered one of two free-response items. They then chose three questions from seven, each of which covered a different topic: industrial chemistry, polymers, electrochemistry, sugars, proteins, molecules and cells in the immune system, or instrumental chemistry (instrumentation).

Japan (1991, 1992)

Entrance Examination, Tokyo University. This examination, allotted 3 hours, consisted of three questions broken into many subquestions. A wide variety of item types were employed.

Sweden (1991, 1992)

National Examination. This 20-question examination was allotted 2.5 hours and had three sections.

• The first section had 14 questions (1 point each) comprised of nine multiple-choice questions and five word/phrase questions.

• The second section had two free-response problems (2-points each).

• The third section had four free-response problems (3-points each).

United States (1989, 1993)

Advanced Placement Examination. This 3-hour examination consisted of two papers.

• Paper 1 was allotted 1.5 hours and had 75 multiple-choice questions. This paper constituted 45 percent of the examination's total score.

• Paper 2 was allotted 1.5 hours and accounted for 55 percent of the total score. It had four sections.

  - The first section was a free-response question with several required subquestions.
  - The second section consisted of two free-response questions with subquestions; students chose one of the two questions to answer.
  - In the third section, students were instructed to answer five of eight free-response questions.
  - In the fourth section, students answered three out of five questions.

General Structure

Table 4-1 summarizes basic structural features of the various countries' examinations for the years included in this study.

Length and Scorable Events

The time required to sit for the chemistry examinations ranged from less than 3 hours to more than 8, and the number of scorable events ranged from 11

Table 4-1.   General Structure of the Chemistry Examinations
Number of Scorable Events (SEs)

Country	Exam Length (Hrs)	Total No. of SEs	Multiple-Choice (SEs)	Free-Response (SEs)	Laboratory Practical (SEs)
E&W	8.3	184	40	114	28
France*	1.4	11	—	11	—
Germany	4.0	41	—	41	—
Israel	5.0	119	16	103	—
Japan	2.5	38	1	38	—
Sweden	3.0	20	8	12	—
US	3.0	114	75	39	—

*   Chemistry questions comprise two of five sections in combined chemistry/physics examinations.

(France) to almost 200 (England and Wales).[1] The US examination was one of the shortest, but contained one of the largest number of scorable events. This suggests that, on average, the scorable events on the US examination must have been much smaller bits of subject matter than on the other examinations.

Choice

Some of the examinations allowed students to chose among certain questions. The English/Welsh, Israeli, and US examinations all had several item sets where students could make choices. On all the other examinations, students were expected to complete every item.

---

[1] Scorable events were the smallest, discrete questions in each examination. They were the unit of analysis in this study and are described in the Approach to Comparing Examinations section of Chapter 1.

Item Characteristics

This section compares item types across examinations and discusses the use of diagrams, photographs, graphs, and tables in these items.

Item Types

The amount of each examination allocated to different item types varied greatly among the set of examinations. This important examination characteristic is described three ways. Table 4-1 already provided the number of scorable events found of each main item type. Similarly, Figure 4-1 shows the amount of each main item type but in terms of how much examination time was allotted to it.

Figure 4-1.   General Item Types in Chemistry Examinations

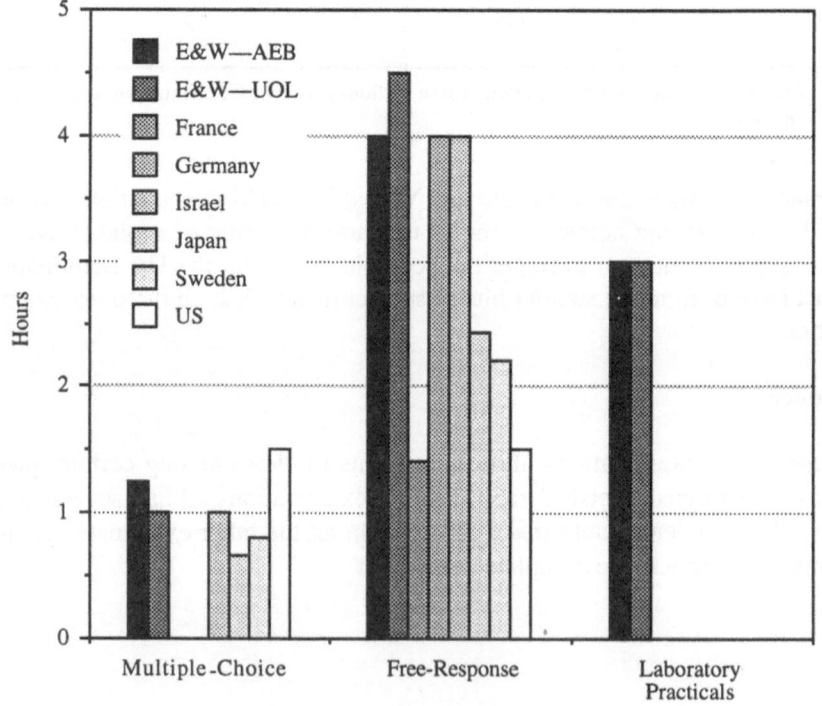

Finally, Table 4-2 shows the proportions of the total examination score accounted for by various item types, and it divides free-response items into short- and extended-answer categories. In every country's examinations, except those of Sweden and the United States, the majority of scorable events were free-response, short-answer items. Only the United States relied on multiple-choice questions for as many as half of the scorable events on its chemistry examinations.

Table 4-2.   Item Type in Chemistry Examinations
Percentage of Examination Score

Country	Multiple-Choice	Free-Response, Short-Answer	Free-Response, Extended-Answer	Laboratory Practical
E&W	16	62	12	**10**
France	—	**83**	18	—
Germany	—	70	**30**	—
Israel	20	67	13	—
Japan	2	**87**	11	—
Sweden	28	50	22	—
US	**50**	44	6	—

Bold indicates highest value(s) in each column.

Multiple-Choice Items. The United States and Sweden were the major users of multiple-choice items, with these items representing 50 percent and 28 percent of the respective examination's scores. England and Wales and Israel also included a substantial number of multiple-choice items (18 and 20 percent, respectively). In every country, the multiple-choice items consisted of four or five responses, only one of which was correct. The French and German examinations included no multiple-choice items, and the Japanese examination included very few.

The debate as to whether some item formats are superior rages in discussions among chemistry educators throughout the world. At the risk of making a simplistic overgeneralization, our impression is that British chemistry educators appeared to be the most antithetical to multiple-choice examinations, and the Americans seemed to be the strongest champions of that format. Attitudes in the other countries appeared to be somewhere in between, or predisposed to ignore the controversy. Interestingly enough, however, one paper in each English/Welsh examination was in a multiple-choice format—and one of the two papers in each year of the US examination was entirely in a free-response format.

Free-Response Items: Short and Extended Answers. Short-answer, free-response items were, by far, the most common type of scorable event on all examinations surveyed except for the United States. Generally, the more difficult extended-answer questions were most extensively used in the German examination, where they accounted for fully 30 percent of the total score. In all other countries—except for the United States—extended-answer items accounted for at least 10 percent of the total examination score.

A significant number of scorable events on the examinations depended on successfully completing an earlier task that was often quite simple. This "building" effect occurred in all of the analyzed examinations except the one from Sweden. On several examinations, this effect accounted for as much as 10 to 15 percent of all free-response items.

Free-Response Items: Word-Phrase Answers. The UOL and AEB examinations contained more items requiring only one-word (or one-phrase) responses than did any other country's examination. Even here, however, these items represented less than 10 percent of the total. Germany, Israel, Japan, and Sweden each had only one to three word-phrase items each: these represented a few percent of the examinations' scorable events. France and the United States did not have any word-phrase items.

Essay Items. Of the examinations surveyed, only the UOL examinations contained an essay questions. Students taking these UOL examinations wrote one 40-minute essay by choosing from four questions. A second, 20-minute essay was required.

Practical Activity Items. Practical tasks were, at one time, a mainstay in chemistry achievement assessments. The laboratory practical was a standard kind of instrument, and was used extensively in awarding marks. For example, a generation ago, practical items were heavily used on published US examinations. Practical activities, however, are used relatively little today. In the examinations surveyed, practical assessment tasks appeared only in the English/Welsh examinations, and represent approximately 10 percent of the point value of these examinations. (See Example 4-1.)

Example 4-1.   (England and Wales, University of London, 1991)
                      Full Question

2.   You are provided with two compounds C and D, and with a compound E, which contains carbon, hydrogen, and oxygen only. Carry out the following tests on C, D, and E and record your observations and inferences for each. Finally, answer the questions which follow the tables. [Tables with spaces for answers are provided.]

a.   To three-quarters of your sample of C in a boiling tube add 10 cm^3 of dilute hydrochloric acid. Heat to boiling point and mix the contents of the tube well. Cool, and filter if necessary. Carry out the following tests on portions of the solution:
     (i)   To 2 cm^3 of the solution of C add aqueous sodium hydroxide.
     (ii) To 2 cm^3 of the solution of C add dilute sulphuric acid.

b.   Carry out a flame test on a little of substance C.

c.   Divide the remainder of substance C into two equal portions. Reserve one portion for test (d) and treat the other portion as follows: add 7-8 cm^3 of distilled water to the portion of C and shake the mixture well. Filter, and divide the filtrate into two equal portions. Perform the following tests on these portions:
     (i)   Test one portion of the filtrate with red and with blue litmus paper.
     (ii) To 1 cm^3 of acqueous magnesium chloride add the other portion of the filtrate.

d.   Mix the reserved portion of C with half of your sample of D. Add 2 or 3 drops of water to this mixture in a test tube and warm the mixture. Test any gas evolved.

e.   Heat very little of E on an inverted crucible lid and ignite E from above.

f.   Divide the remainder of E into two roughly equal portions. reserve one portion for test (g) and use the other portion as follows: Shake the portion of E with 6-7 cm^3 of distilled water and test the mixture with red and with blue litmus paper.

g.   To the reserved portion of E in a boiling tube add 2 cm^3 of methanol and 2 drops of concentrated sulphuric acid (CARE). Carefully heat the mixture for one minute in a boiling-water bath. Pour it into a little aqueous sodium carbonate in an evaporating dish.

h.   The anion in D is a halide. Describe a procedure which would enable you to deduce the identity of this anion. Carry out the procedure and record your observations and inferences below.

Now answer the following:
(1) Suggest an identity for C.
(2) Suggest an identity for D.
(3) Sketch below the structural features of E.

Use of Diagrams, Graphs, and Table

Most countries made little use of tabulated or graphical elements—less than one might expect given the learning differences among students and the nature of chemical research and industry. (See Figure 4-2.) This lack is in sharp contrast to the actual practice of chemistry, where chemists and chemical engineers routinely construct graphical, tabular, or diagrammatic displays to convey the substance of their work. Further, items that include drawings or photographs can substitute for extensive verbiage, decreasing the extent to which a test assesses reading ability. Examinations relying predominantly on text to convey information or data benefit students who can deal with that format while penalizing students who cannot.

Figure 4-2.   Use of Diagrams, Graphs and Tables in Chemistry Examinations

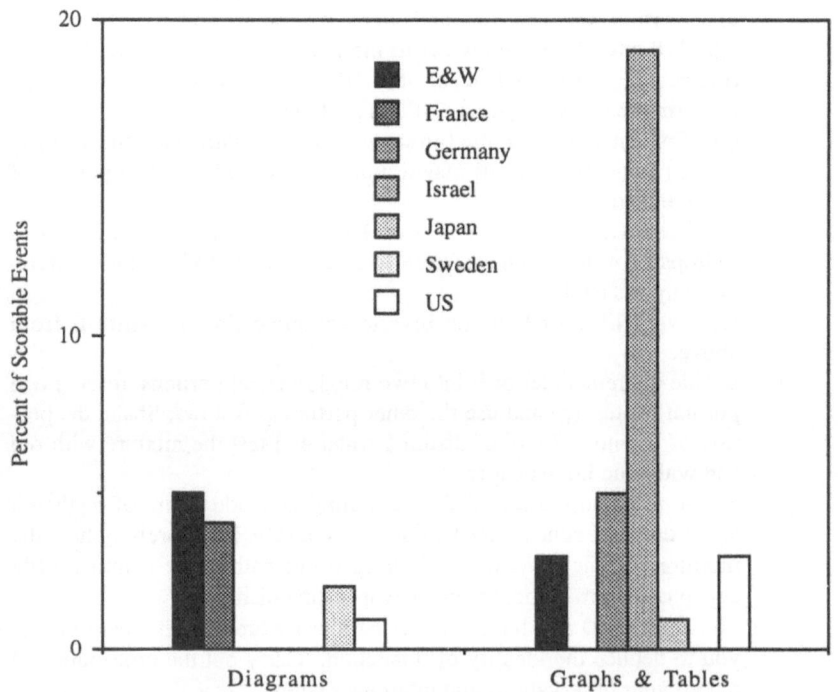

The Israeli examination made the most extensive use of tables and graphs, but included no diagrams. The English/Welsh examinations, on the whole, included the most varied use of diagrams, tables, and graphs; but only 8 percent of these examinations' total scorable events contained one of these elements. Examinations from the other five countries made even more limited use of diagrams, graphs, and tables.

Examination Topics

This section compares the examinations' topics in two complementary ways. The first subsection notes contrasts only among the most emphasized topics of each examination while the second subsection indicates all chemistry topics covered by each examination. The latter, comprehensive reporting of all topics across all examinations focuses on general topics, such as organic chemistry. The first comparison, dealing with the examinations' top ten topics, looks at more detailed topics such as mechanisms of organic reactions, addition and substitution reactions, or hydrocarbons.[2] In addition to providing contrasts among all chemistry topics, this section gives prominent attention to marked differences in how the examinations dealt with (or failed to deal with) organic chemistry, redox reactions, environmental chemistry, and industrial applications.

Emphasized, Detailed Topics in Each Examination

Overview of Most Emphasized Topics. Table 4-3 indicates the most common, detailed topics for each of examinations in this study.[3] Oxidation and reduction reactions was the most frequent topic, appearing in every country's examinations. Acids and bases, equilibrium expressions, and the mole concept appeared in every country except one; England and Wales omitted equilibrium expressions and the mole concept while Germany did not include acids and bases. Three topics were included in all but two countries' examinations: addition/substitution reactions, formulas/equations, and organic oxygen and nitrogen compounds.

---

[2] Appendix A may be helpful for understanding the meaning of the more general topics because it lists the specific topics associated with them. General topics were taken from the Science Framework of the Third International Mathematics and Science Study (Robitaille et al., 1993). The detailed topics were added to permit more specific descriptions of the examinations.

[3] The ten most-emphasized topics are noted for each examination. The table lists the most frequently occurring topics first.

Table 4-3.   Chemistry Topics, Detailed: Ten Most Emphasized in Each Examination

Percentage of Examination Score

Topic	E&W	France	Germany	Israel	Japan	Sweden	US	Topic Frequency*
Redox	7	4	12	9	6	12	7	7
Acids and bases	7	9		9	5	12	3	6
Equilibrium		31	6	8	6	15	14	6
Mole concept		4	2	6	9	20	17	6
Addition/ substitution	5	9	7		9	3		5
Formulas/ equations	7	4	13		9	20		5
Organic O, N compounds	5	5	8	5	15			5
Second law	5			12			6	3
Covalent compounds	4					2	5	3
Periodicity	5						4	2
Catalysis	4				7			2
Organic mechanisms		4	6					2
Biochemistry				7	11			2
Ionic reactions					11		4	2
Solutions				5			4	2
Rate laws		21						1
Polymers			7					1
Evidence of change	6							1
Biology					5			1

Table 4-3.   Continued

Topic	E&W	France	Germany	Israel	Japan	Sweden	US	Topic Frequency*
Free energy				5				1
Phase changes					5			1
Inter-particle forces		4						1
Hydrocarbons						3		1
Galvanic cells						3		1
Electron configuration							3	1
Ionic compounds						2		1
Mass/volume		2						1
Total - Topic Concentration **	32	75	47	49	53	79	49	

Bold indicates values of 10 percent or greater.
Blank cells only indicate that topic was not one of the top ten in the examination. The examination still may have included the topic.
*   Topic frequency indicates the number of countries that emphasized the topic.
** Topic concentration indicates the percentage of the examination comprised by its ten most emphasized topics.

The remaining topics in Table 4-3 were a second-order presence among the examination set, appearing as emphasized topics in only one, two, or three countries' examinations. While most of these topics only constituted 2-7 percent of any country's examinations, four topics did occur as more than 10 percent of a single country's examinations: rate laws (21 percent, France), the second law of thermodynamics (12 percent, Israel), biochemistry (11 percent, Israel), and ionic reactions (11 percent, Japan). Some topics that are generally regarded as very elementary for advanced study—such as electron configuration and some aspects of periodicity—were found almost exclusively in the US examinations.

Using a broad definition of organic chemistry—organic oxygen and nitrogen compounds, biochemistry, polymers, biology, and hydrocarbons—

these examinations allocated considerable coverage to organic chemistry topics. The glaring exception was the United States, where only 1 percent of the AP examinations addressed organic chemistry topics, a number much too low to appear in its emphasized topic list. All five organic subjects appeared in German examinations, three were found in French and Israeli examinations, two were found in the Japanese and Swedish examinations, and one topic was emphasized in England and Wales. Additionally, addition/substitution reactions—which primarily are carbon-based—appeared on most examinations. The Swedish examination, however, was the only one that covered hydrocarbons so extensively that this topic appeared on its list of emphasized topics.

Topic Concentration. The summary row of Table 4-3 reveals the examinations' topic concentration. It shows the percentage of each country's examinations comprised by its 10 most emphasized topics.[4] At the extremes, examinations in Sweden and France focused tightly on a few topics, while those in England and Wales covered a broad spectrum in chemistry.

More specifically, the Swedish top ten topics comprised the majority (79 percent) of the examinations. The high percentage (75 percent) for French chemistry examinations reflected their structure. Since French chemistry examinations were allotted only 1.2 hours and contained 11 questions, the most common topics must constitute the bulk of the examination. In England and Wales, only 32 percent of the examinations' topics were accounted for by the 10 most emphasized topics; Conversely, the majority of these examinations (68 percent) were devoted to many other topics. For Germany, Israel, Japan, and the United States, the 10 most emphasized topics represented about half (47-53 percent) of the examinations.

Varied Treatments of Redox Equations. To close, this section presents typical redox scorable events in an effort to illuminate the contrasting styles of setting questions for students to handle. Again, every country included scorable events that required students to do something with redox reactions. By far, the most common oxidizing agents involved manganese or chromium compounds. (All seven countries included compounds of one of these two elements somewhere in their examinations.) Some of the scorable events dealt with redox equations either isolated from, or only casually linked to, application or implication. The United States, English/Welsh, Israeli, and Swedish scorable events fit this model (see Examples 4-2

---

[4] This statistic can be thought of as an indicator of an examination's breadth in topic coverage. Examinations with higher and lower topic concentrations focused on fewer or more topics, respectively.

through 4-5). In the Swedish and English/Welsh scorable events, students were asked to construct the redox equation (or half-equation) themselves. The Israeli scorable event was somewhat simpler, in that students only had to combine half-equations that were supplied to produce a net equation. The US scorable event was even simpler: students were only asked to identify the species undergoing oxidation or reduction.

Example 4-2.   (Sweden, 1991)
Full Question

If you add sulfur dioxide to an acidic solution of potassium dichromate, $K_2Cr_2O_7$, the sulfur dioxide is oxidized to sulfate ion and the dichromate ions are reduced to chromium(III) ions. Write the equation with the smallest possible whole number coefficients. Write free ions in water solution as free ions.

Example 4-3.   (England/Wales, University of London, 1992)
Partial Question

Permanganate ion is a very powerful oxidizing agent used in quantitative volummetric analysis. In many cases, the reaction is slow to begin, but once established, proceeds rapidly.

(i) Write a half-equation to show how permanganate ion behaves as an oxidizing agent in acidic solution.

Questions ii and iii follow.

Example 4-4.   (Israel, 1992)
Partial Question

All alkaline cell is based on the following two half-reactions:

$$2MnO_2(s) + H_2O(l) + 2e^- \rightleftarrows Mn_2O_3(s) + 2OH^- (aq)   E^\circ = -0.15 \text{ V}$$

$$Zn(OH)_2(s) + 2e^- \rightleftarrows Zn(s) + 2OH^- (aq)   E^\circ = -1.25 \text{ V}$$

a. Write the equation for the overall reaction that occurs in the cell.

Questions b-e follow.

Example 4-5. **(United States, 1989)**
Full Question

$6I^- + 2MnO_4^- + 4H_2O(l) \rightarrow 3I_2\ (s) + 2MnO_2(s) + 8OH^-\ (aq)$

Which of the following statements regarding the reaction represented by the equation above is correct?
(A)   Iodide ion is oxidized by hydroxide ion.
(B) MnO_4^- is oxidized by the iodide ion.
(C)   The oxidation number of manganese changes from +7 to +2.
(D)   The oxidation number of manganese remains the same.
(E)   The oxidation number of iodine changes from -1 to 0.

The German scorable event (Example 4-6) cast the redox problem in terms of an industrial application, and asked students to complete stoichiometric calculations based on a redox titration. They wrote and balanced the required equations themselves, as well as described the experimental process.

Example 4-6. **(Germany, Bavaria, 1991)**
Partial Question

The waste water of a bleaching facility is manganometrically tested for the hydrogen peroxide content. In the test, a 200-mL sample of the waste water consumes 37.5 mL of a 0.1 M potassium permanganate solution.

1.1   Describe the experimental processes involved in this volummetric analysis.
1.2   Derive the net reaction equation from the half-equations.

Question 1.3 follows.

The Japanese redox scorable event (Example 4-7) had much of the same feel as the German scorable event, except that it was part of a much larger string of interrelated scorable events based on analyzing mixtures that called for concepts and knowledge from various branches of chemistry.

The French redox scorable event (Example 4-8) involved oxidation of an organic compound, and focused more on the organic compounds than on the processes of oxidation and reduction. Students nonetheless had to complete and balance the redox equation.

It is clear that the various examinations had similar objectives related to the testing of students' knowledge of the process of oxidation and reduction.

Example 4-7.   (Japan, 1992)
  Partial Question
  Questions F and G precede.

Concentrated hydrochloric acid is added to a mixture containing manganese dioxide and inert ingredients. 0.67 L of chlorine gas (measured at 0°C and 1 atm pressure) is produced.

H. Find the number of moles of manganese dioxide (to two significant figures) in the original mixture.

Example 4-8.   (France, Aix, 1992)
  Full Question

For the reaction presented below, first find the structural formula of the compound B; then write, and balance if necessary, the equation and name each chemical species.

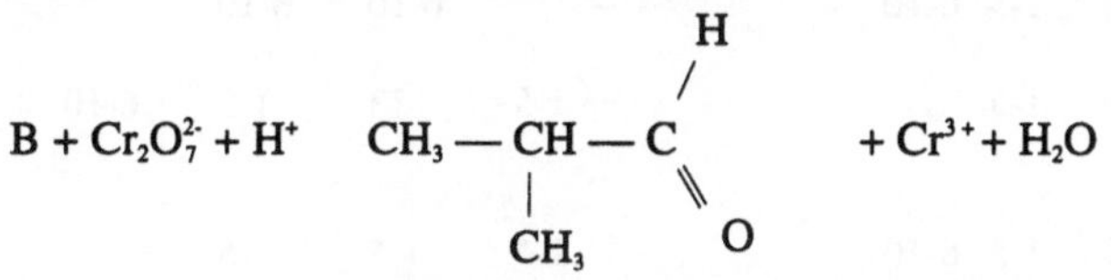

$$B + Cr_2O_7^{2-} + H^+ \quad CH_3 - CH - C\begin{matrix} \diagup H \\ \diagdown \diagdown O \end{matrix} \quad + Cr^{3+} + H_2O$$
$$\qquad\qquad\qquad\qquad | \\ \qquad\qquad\qquad CH_3$$

The variation in how those objectives translated into actual scorable events is quite remarkable, entailing differing levels of complexity and sophistication.

All General Topics Covered by Each Examination

Overview of Topic Coverage. Table 4-4 compares how all general topics were covered by the examinations.[5] Not surprisingly, the content of the various countries' examinations was fairly constant between examination boards. This finding is to be expected if the examinations do in fact reflect curricular objectives within the secondary schools of those countries.

The following topics were well-represented on most of the examinations: chemical change; atoms, ions, and molecules; and reaction rates and equilibria. Other expected topics such as energy, electrochemistry, and subatomic particles also appeared in most of the examinations, but in varying

---

[5] Those topics which received the greatest emphasis appear first, with categories receiving less emphasis following in descending order.

Table 4-4.   Chemistry Topics, General: All Topics Found in Each Examination
Percentage of Examination Score

Topic	AEB	UOL	Aix	Paris	Bav	B-W	Israel	Japan	Sweden	US
Chemical change	18	20	18	6-10	12	19	21	30	23	17
Organic/ biochemistry	17	11	15	25	35	21	12	43	6-10	1-5
Atoms, ions, molecules	19	22	6-10	6-10	22	17	25	1-5	44	29
Rates/ equilibrium	6-10	6-10	50	59	6-10	6-10	6-10	13	15	19
Physical change	6-10	6-10	1-5		1-5		14			6-10
Energy	1-5	6-10				6-10	6-10		1-5	6-10
Macro-molecules	1-5	1-5			1-5	13	1-5	6-10	1-5	
Subatomic particles	1-5	6-10			1-5	1-5	1-5		1-5	1-5
Electro-chemistry	1-5	1-5				6-10	1-5		1-5	6-10
Chemical properties	6-10	6-10	1-5				1-5			
Explanation/ physical	1-5	1-5			1-5	1-5	1-5			1-5
Nuclear chemistry	1-5	1-5			6-10					1-5
Physical properties	1-5	1-5	1-5				1-5			1-5
Explanation/ chemical	1-5	1-5								
Kinetic theory	1-5									1-5
Classification					1-5					

Table 4-4.   Continued

Topic	AEB	UOL	Aix	Paris	Bav	B-W	Israel	Japan	Sweden	US
Quantum theory										
Environmental chemistry										
Industrial chemistry	6-10		1-5		**14**			6-10	6-10	

Bold indicates values greater than 10 percent. Table values 1-5 and 6-10 indicate that averages for two years of these examinations fell within these ranges. Values less than 0.5 percent are not included.

proportions. Virtually all of the scorable events on all of the examinations fit very definitely into the "traditional" chemistry topics. The only exceptions involved the Israeli examination—which included some straight biology items—and one English/Welsh examination, which included some practical tasks that, in our judgment, did not contain chemistry content.

Contrasts in Topic Coverage Between US and Other Countries. The US examinations devoted their greatest attention to atoms, ions, and molecules (29 percent); reaction rates and equilibria (19 percent); chemical change (17 percent); and physical change (10 percent). Together, these four topic groups comprised 74 percent of the US examinations, as seen in Figure 4-3. While Sweden, France, and Japan devoted comparable attention to these topic groups, the distribution among the topics varied sharply. These four topic groups were introductory subjects in chemistry curricula almost everywhere. The fact that relatively much less attention was devoted to them in the English/Welsh, German, and Israeli examinations seems to be an indicator that the curricula forming the base for these examinations have moved beyond introductory topics, and examiners no longer feel a compunction to address such elementary matters.

One of the first subjects addressed in introductory chemistry courses is atomic structure. Beginning students must have a reasonably thorough knowledge of electron configuration before they move on to—among other topics—chemical bonding and molecular structure. In general, the examinations did not cover this fundamental topic: none contained any atomic structure questions. It can only be concluded that advanced secondary-level students in these countries were so far past this topic that it no longer appeared on their examinations. The only instances of electronic structure scorable events on the examinations surveyed were on the English/Welsh and US

Figure 4-3.   Emphasis Given to Topic Groups That Dominate US Examinations

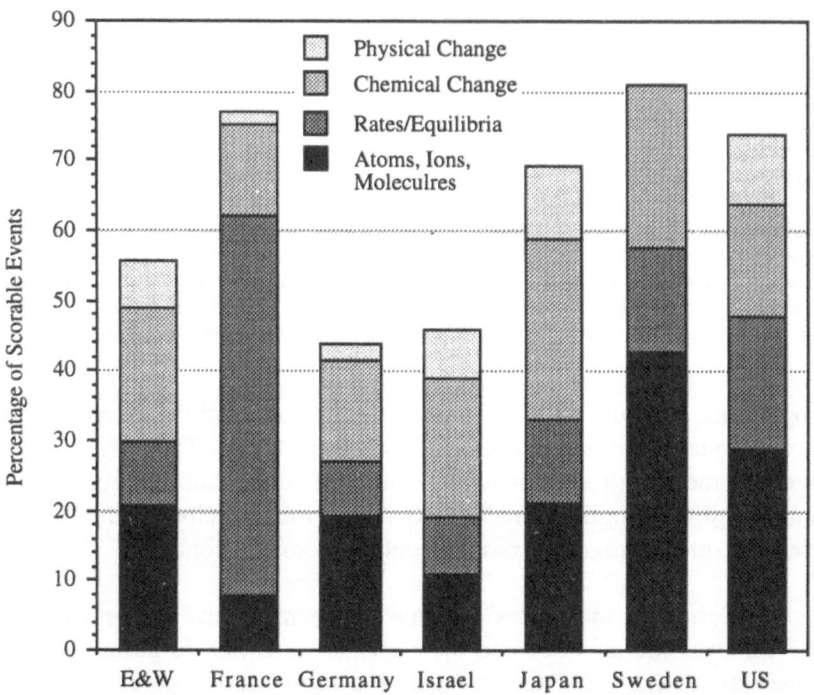

examinations, as illustrated by Examples 4-9 and 4-10. In both cases, the scorable events were cast in multiple-choice format and required about the same level of knowledge.

Example 4-9.   (England/Wales, University of London, 1991)
Full Question

Which of the following electronic configurations represent(s) an atom of a *d*-block element?

1	$1s^2 2s^2 2p^6 3s^2 3p^6 3d^5 4s^2$
2	$1s^2 2s^2 2p^6 3s^2 3p^6 3d^{10} 4s^1$
3	$1s^2 2s^2 2p^6 3s^2 3p^6 3d^{10} 4s^2 4p^1$

(A)   1, 2, and 3 are correct.
(B)   1 and 2 only are correct.
(C)   2 and 3 only are correct.
(D)   1 only is correct.
(E)   3 only is correct.

Example 4-10.   (United States, 1993)
Full Question

$$1s^2 2s^2 2p^6 3s^2 3p^6 3d^2$$

In which of the following pairs do both species have the ground-state electronic configuration shown above?

(A) Ti^{2+} and V^{3+}
(B)   Sc and Ca^{2+}
(C) Sc^+ and Mn^{3+}
(D)   V and Mn^{3+}
(E) Ca^{2+} and Ti

The feature that most distinguished the US examination from those of virtually every other country was a lack of attention to organic chemistry and biochemistry, and to macromolecules. Figure 4-4 shows that on all but the US examinations, as much as 30 to 40 percent of the examination questions were drawn from these two topic groups. Virtually every country except the United States expected students to demonstrate substantial understanding of the chemistry of organic compounds, biologically important compounds, and/or macromolecules. These topics are treated as advanced subjects that first receive serious attention no earlier than the sophomore college course in US chemistry curricula. Chemistry curricula in most other industrialized nations address these topics much earlier, although still following appropriate grounding in fundamental concepts and facts.

The German examinations, in particular, devoted more than 35 percent of the points on their examinations to organic chemistry and biochemistry and to polymer chemistry. This great attention to the chemistry of carbon compounds was congruent with the enormous contributions made to organic chemistry by German chemists during the 19th and early 20th century. The historical attention to organic chemistry among a country's scientists cannot be the only factor, however. Every other country surveyed, including those that have a rather modest history of contributions to the field, included substantial attention to carbon-based chemistry.

The emphasis of US chemistry curricula on physical chemistry topics, perhaps reflects the residual influence of the *ChemStudy* project of the 1960s. The corpus of material that defines US chemistry courses at the secondary level has changed little since the curriculum revolution of the post-Sputnik era when NSF-funded *ChemStudy* became established as the *de facto* higher-level standard for high school chemistry. The influence of *ChemStudy* extended far beyond the first year of high school chemistry. Introductory col-

Figure 4-4.   Emphasis Given to Organic and Biochemistry Topics

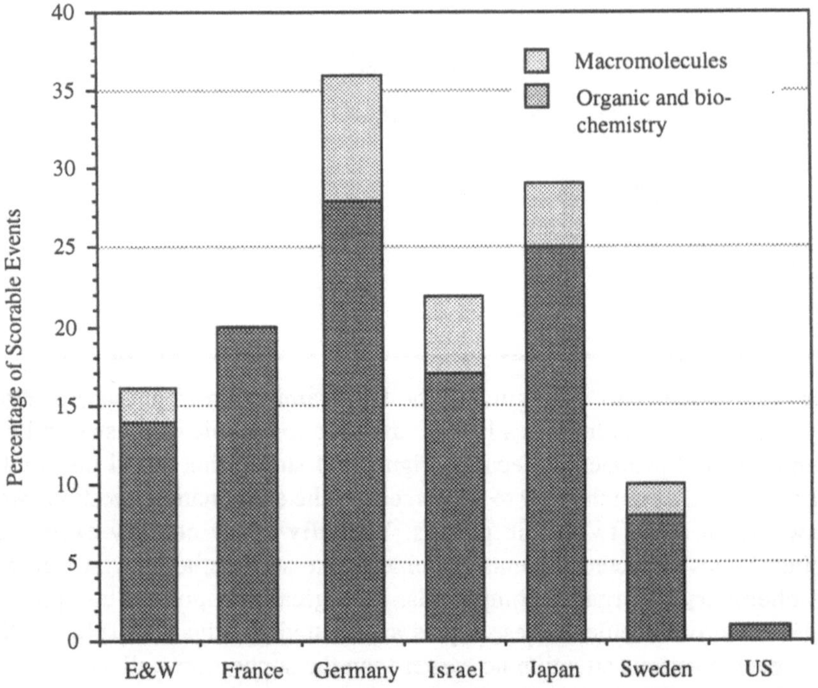

lege textbooks, which are also widely used for second-year high school courses, were re-written with the intention of coordinating with *ChemStudy*. Advanced Placement courses, which usually constitute a second year of high school chemistry, reflect the topic coverage of most general chemistry courses at the college-level with considerable fidelity.

The central fact of modern industrial chemistry and the frontiers of chemical research is that more than half of the world's practicing chemists devote their professional lives to one aspect or another of the chemistry of carbon compounds. The United States appears to be the only country that fails to include significant attention to the field in curricula for secondary students.

Varied Treatments of Organic Chemistry. The following sample examination questions illustrate typical treatments of organic chemistry on the various examinations. Examples 4-11 and 4-12 illustrate the handling of organic chemistry in England and Wales and France.

Example 4-11.   (England/Wales, Associated Examining Board, 1991)
             Full Question

Benzene can be nitrated by treating it with a mixture of concentrated nitric acid
and sulfuric acid. After separation from the reaction mixture, the nitrobenzene
formed can be purified by steam distillation.
(i)   What type of mechanism is involved in nitrations of this type?
        Give the name of the nitrating species and write equations to show how
        it is formed in the nitrating mixture.
(ii)   What reaction conditions are necessary to ensure a good yield of nitro-
        benzene? Explain your answer.
(iii)   Outline the technique and theory of steam distillation.

Example 4-12.   (France, Paris, 1991)
             Full Question

Three pairs of organic compounds are presented in this problem. In each pair,
the two compounds are isomers.
(1)   Indicate the chemical function and name of each compound.
(2)   Describe the experimental test for differentiating the two compounds of
        pairs I and II.

A Bunsen burner and the following reagents are available for the experiment:
dinitrophenylhydrazine, potassium dichromate in an acidified solution,
Fehling's solution, Tollen's reagent (containing the diamminesilver(I) ion).

Pair I $CH_3 - C - CH_3$ $CH_3 - CH_2 - C \overset{\diagup H}{\underset{\diagdown O}{}}$
                      $\parallel$
                      O

Pair II $CH_3 - \overset{\displaystyle OH}{\underset{\displaystyle CH_3}{\overset{\displaystyle |}{\underset{\displaystyle |}{C}}}} - CH_3$ $CH_3 - CH_2 - CH_2 - CH_2OH$

Pair III $CH_3 - CH_3 - C \overset{\diagup O}{\underset{\diagdown OH}{}}$ $CH_3 - C \overset{\diagup O}{\underset{\diagdown OCH_3}{}}$

This question style and item content were not unique. Consider similar items from the German, Israeli, and Japanese examinations. (See Examples 4-13, 4-14, 4-15)

Example 4-13.   (Germany, Bavaria, 1992)
            Full Question

The compounds ethane, monochloroethane, and diethyl ether can all be derived from a compound having the molecular formula C_2H_6O by way of a common intermediate product.

3.1   Using structural formulas, give the first reaction step to the intermediate product which is common in the formation of all three of the molecules.

3.2   Using structural formulas, write the reactions leading from the intermediate to each of the compounds ethane, monochloroethane, and diethyl ether.

Example 4-14.   (Israel, 1992)
            Full Question

In quantitative analysis of compounds A and B, each with a molar mass of 60 g•mol⁻¹, they have been found to have the same composition: 60 percent of the mass of each is carbon; 13.34 percent is hydrogen; and the remainder is oxygen.

a. Determine the molecular formula for compounds A and B. Show your calculations.
b. Draw 3 possible structural formulas that fit the molecular formula that you derived.

The data in the following table relate to compounds A and B and to an additional compound C, that is also composed of carbon, hydrogen, and oxygen.

Reaction w/ acidic $K_2Cr_2O_7$	Reaction w/ compound acidic $ZnCl_2$	Solubility in water	Boiling pt (°C)	Molar mass (g•mol⁻¹)	compound
+	only after extensive heating	+	97.1	60	A
-	-	+	7.0	60	B
+	-	+	48.8	58	C

c. Draw the structural formulas for A and B.
d. Based on the data in the table, determine which is the functional group in compound C. Explain.

Example 4-15.   (Japan, 1991)
Full Question

Read sentences (a) - (e) concerning an ester and answer the questions that follow. Use these atomic weights and write the empirical formula:
C=12.0; H=1.0; O=16.0.
(a)   The ester has a molecular formula $C_{11}H_{14}O_2$.
(b)   An alcohol and a carboxylic acid are obtained from the hydrolysis of the ester.
(c)   The alcohol is 64.9% C; 13.5% H; and 21.6% O. It oxidizes to an aldehyde that has an identical mass of carbon as the alcohol.
(d)   The carboxylic acid is aromatic.
(e)   A molecule of an organic compound generally is three-dimensional. In terms of both carbon and oxygen atoms within a molecule, however, all the carbon and oxygen atoms can be lined up in the same plane for butane and ethyl acetate. On the other hand, all carbon and oxygen atoms cannot be lined up in the same plane for this ester.

(1)   What is the molecular formula for the alcohol? How many isomers are there, including optical isomers, that have an identical molecular formula?
(2)   When the maximum number of carbon and oxygen atoms line up in the same molecular plane for the ester, which carbon atom (or atoms) is (or are) out of the plane? Write the structural formula for the ester and circle the out-of-plane carbon atoms.
(3)   Explain why the carbon atom(s) identified above must be out of the plane of the other carbon atoms.
(4)   The alcohol has isomers that are ethers and has the identical molecular formula. Of these, write the structural formulas of the isomer which cannot have all the carbon atoms and the oxygen atom in the same plane.

The contrast between the preceding questions and questions from the Swedish and US examinations is enormous. (See Examples 4-16, 4-17.)

Example 4-16.   (Sweden, 1991)
Full Question

Which three of the following compounds have the formula C_3H_6O?
(A)   Propanal
(B)   1-Propanol
(C)   Propanone
(D)   Propanoic acid
(E)   2-Propene-1-ol

Example 4-17.   (United States, 1991)
Full Question

Which is used to explain the fact that the carbon-to-carbon bonds in benzene, C_6H_6, are identical?
(A)   Hydrogen bonding
(B)   Hybridization
(C)   Ionic bonding
(D)   Resonance
(E)   van der Waals forces (London dispersion forces)

Little Treatment of Industrial Chemistry. [6] Figure 4-5 illustrates that only the examinations of Sweden and the United States contained no industrial chemistry. The German examinations, on average, contained twice as much emphasis on industrial and practical applications of chemistry as those of any other country (see Example 4-6 for a sample question having an industrial context).

Figure 4-5.   Emphasis Given to Industrial Chemistry Topics

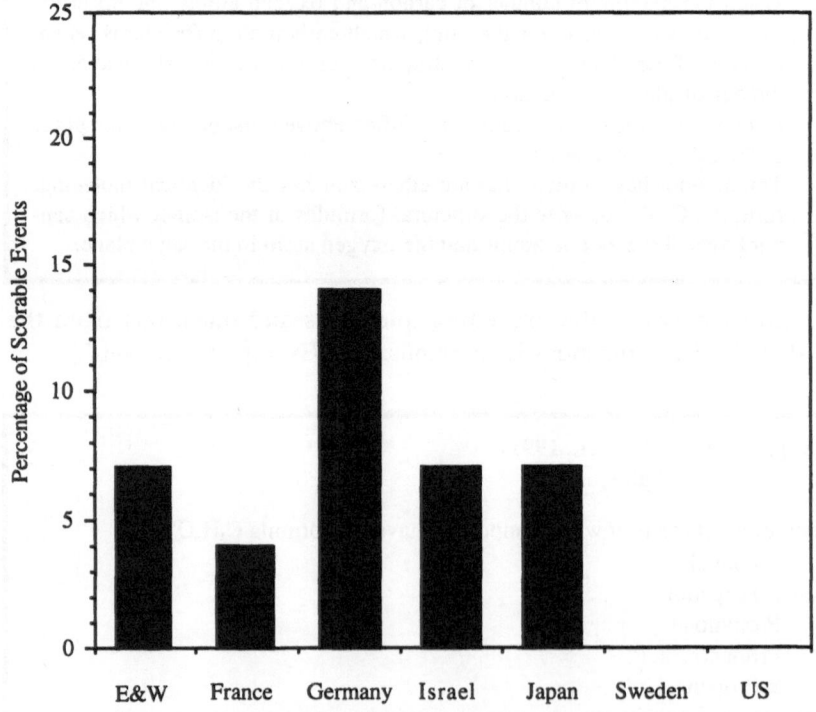

The Bavarian emphasis on industrial chemistry was certainly congruent with German chemists' historical position as a world leader in industrial organic chemistry. On the other hand, as the chemical industry in the United States emerged as the largest and most diverse producer of organic chemicals and of finished goods produced from organic intermediates, one would have expected the American educational system to begin to mirror this change. Such has clearly not occurred. Just as the US secondary schools' chemistry curricula ignored organic chemistry topics, these curricula also ignored industrial applications altogether. The American examinations reflected this. Example 4-18 is an example of an Israeli question in industrial chemistry.

Example 4-18.   (Israel, 1991)
                   Partial Question

The following question deals with methods for production of $NaBrO_{3(s)}$
Following is balanced formula for the reaction:

I. $7Na_{(aq)} + Br_{(aq)}^- + 6OH_{(aq)}^- + 3Cl_{2(g)} \Leftrightarrow 7Na_{(aq)}^+ + BrO_{3(aq)}^- + 3H_2O_{(L)}$

   Questions a-d precede.

In the bromine plants in Israel, a different method is used to produce $NaBrO_{3(s)}$:

II. $3Br_{2(L)} + 6Na_{(aq)}^+ + 6OH_{(aq)}^- \Leftrightarrow 6Na_{(aq)}^+ + BrO_{3(aq)}^- + 5Br_{(aq)}^-$

e.   Write down <u>one</u> industrial advantage of the method described in Formula I over the method described in Formula II.

f.   What was the consideration that led to the choice of method II in Israel? <u>Explain</u>.

g.   How is the percent <u>yield</u> increased in Israel?

Little Treatment of Environmental Chemistry. There was almost no consideration of environmental chemistry topics on any of the examinations. The extent to which any scorable event could be construed to have environmental implications occurred so infrequently that we were unable to effectively code for this subject. Since environmental chemistry is not a core area in the field of chemistry or in general chemistry courses at the college level, it may not be surprising that examinations for college-bound students would omit environmental chemistry. Given the tremendous impact of chemistry on the environment and the resulting impact on daily life, however, it is disappointing that these examinations did not address environmental chemistry.

Performance Expectations

Figure 4-6 illustrates the general categories of performance expectation emphasized by the various chemistry examinations.[7] Perhaps the most interesting observation from the available data is the almost total absence of examination scorable events that dealt with abstracting and deducing scientific principles or making decisions about science. On the other hand, the often largely algorithmic tasks of solving problems, developing explanations, and using models dominated the examinations.

Figure 4-6.   General Performance Expectations in Chemistry Examinations

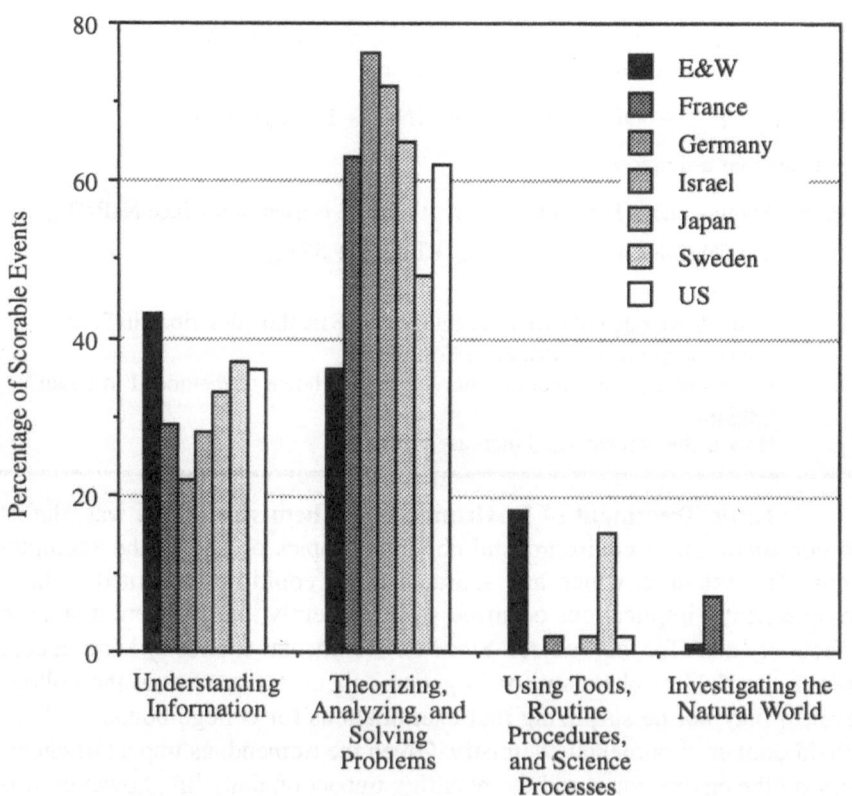

---

[7] Performance Expectations, categories used in the Third International Mathematics and Science Study to describe the student behaviors elicited by various types of questions, are further described in the text accompanying Tables 1-3 and 2-2.

Table 4-5.   Detailed Performance Expectations in Chemistry Examinations
Percentage of Examination Score

Expectation Category	E&W	France	Germany	Israel	Japan	Sweden	US
Understanding							
Simple Information	27	28	12	25	25	20	23
Complex Information	16	2	10	3	8	17	13
Theorizing, Analyzing, and Solving Problems							
Abstraction/ Deduction	—	2	—	—	—	—	—
Qunatitative Problems	7	26	15	16	13	30	33
Developing Explanations	17	14	29	28	20	—	17
Constructing, Using Models	11	21	32	24	24	18	10
Making Decisions	—	—	1	3	9	—	1
Using Tools, Routine Procedures, and Science Processes							
Apparatus, Equipment	—	—	1	—	—	—	—
Routine Experiments	4	—	—	—	—	—	—
Gathering Data	3	—	1	—	—	—	—
Organizing, Representing Data	1	—	—	—	—	—	—
Interpreting Data	9	—	—	—	2	15	2
Investigating the Natural World							
Designing Investigations	1	7	—	—	—	—	—
Conducting Investigations	—	—	—	—	—	—	—
Interpreting Data	—	—	—	—	—	—	—
Formulating Conclusions	—	—	—	—	—	—	—

Differences in performance category assignments across countries were not as striking as differences in content. This is partially because coding assignments were constrained by fewer available categories, and also because Understanding-level scorable events seemed to predominate with both initial assessments covering new material and with cumulative assessments long after instruction had occurred. There were, however, important differences in what students were asked to do within each general performance category. Table 4-5 presents more detailed data on performance expectations in the four broad categories shown in Figure 4-6.

Understanding

On average, about a third of the chemistry examination questions fit into the Understanding performance category. Within that category, distinctions were made as to whether the information dealt with a single concept or fact, or with a more complex array of interrelated concepts or facts. In answering items classified in the Understanding category, students needed to recall—but not process—information.

Solving Problems, Developing Explanations

Tasks in this second group dealt with problem solving, and required the student to manipulate his or her science knowledge to develop new understandings or make decisions. This category accounted, on average, for more than half the questions in the chemistry examinations. Some of the tasks undertaken on items in this performance category involved fairly simple algorithmic tasks, such as constructing formulas and equations. Others required synthesizing information to accomplish the task—sometimes gleaning relationships that had not been addressed in formal study. Thus, items in this group represented a great range of mastery of the material. The examples cited throughout the chapter include fair representations of performance category assignments for the categories of Understanding and Problem-Solving.

Using Tools, Routine Procedures, and Science Processes

These task sets may be "thought experiments" of a "how would you display the relationship between variables x and y" variety, or they may be actual laboratory investigations. It is in this performance expectation group that the real differences among the examinations were revealed, particularly with regard to the English/Welsh and Swedish examinations. Fifteen percent or more of the scorable events of these examinations were classified in this category. On the other countries' examinations, fewer than 2 percent of questions dealt with Using Tools, Procedures, or Processes.

This category does not necessarily require a greater understanding of chemical facts and principles than the preceding category. Both groups of tasks required more of the student than do Understanding tasks, but were at about the same level of difficulty. Since few examination questions addressed Using Tools and Routine Procedures, none of the chapter's previous sample questions illustrate it. However, Example 4-19 deals with using the tools and routine procedures of chemistry to obtain and interpret data.

Example 4-19.   **(Sweden, 1992)**
Full Question

Vinegar is composed of acetic acid, natural flavors, and water. The mass percentage of acetic acid in a certain kind of vinegar was to be determined during a lab. A sample was taken from the vinegar. The mass was found to be 10.4 g. The sample was diluted with about 25 cm^3 of distilled water. A few drops of phenolphthalein was added and the solution was titrated with 0.50 mol/dm^3 sodium hydroxide solution. When 19.3 cm^3 of sodium hydroxide had been added, the indicator changed from colorless to pink color. Determine the percentage of acetic acid in the vinegar.

Investigating the Natural World

The fourth group of tasks also involved using the tools of the chemical sciences, but in much more elaborate ways. Here students were expected to design and conduct investigations, interpret the data they gathered, and formulate conclusions. These are the types of tasks undertaken by research chemists in the real world, and potentially represented the most difficult tasks of all—primarily because success required applying a breadth of knowledge in previously unanticipated ways to reach new understandings.

Only one of the French examinations invoked this performance category, asking students to design an investigation. Examples 4-1 and 4-12 also address—but only marginally—the Investigating the Natural World performance category. However, the AEB question in Example 4-20 did deal with investigating natural phenomena.

Example 4-20.   (England/Wales, Associated Examining Board, 1992)

(Note: This question arises in the course of a laboratory investigation. The student has standardized a sodium hydroxide solution against aminosulfonic acid and has standardized a hydrochloric acid against the sodium hydroxide solution, using the heat evolved to indicate the stoichiometric relationship.)

Comment on the accuracy of this method for determining the concentration of hydrochloric acid. Suggest an improvement of the method that requires no additional apparatus.

Summary

Based on this analysis of chemistry examinations, it is clear that chemistry curricula in these countries were very different—reflecting differences in what chemistry educators in the various countries considered to be appropriate content in the instructional program of upper secondary students. The greatest difference was between the US chemistry curricula and that of the other six nations. Most of these latter had some strong similarities, placing heavy emphasis on organic chemistry, biochemistry, macromolecular chemistry, and industrial chemistry. On the other hand, the US chemistry curricula emphasized physical chemistry topics, perhaps reflecting the residual influence of the *ChemStudy* project of the 1960s, the last time that a major rethink of traditional high school chemistry occurred.

The selected scorable events in this chapter speak volumes about the emphasis, examining styles and expectations of the various examinations. The great differences in the breadth and depth of student education in chemistry among the countries surveyed is clear; the implications of those differences are less clear. If chemical concepts were developed in the same sequence in the curricula of the countries surveyed, secondary students abroad were generally much further along in their study of chemistry than were US students. Items dealing with introductory topics in US curricula were almost entirely lacking in the examinations of most other countries. If, on the other hand, topics were introduced in an altogether different sequence in these other curricula, the difference may not be nearly so great. US chemistry educators generally acknowledge that understanding chemical concepts is not nearly as hierarchical as was once presumed, and more innovation is occurring in US curricula. It was increasingly common, for example, to include organic chemistry topics extensively, even in beginning courses.

The apparent differences in depth are more problematic. In contrast to examinations from countries like England/Wales and Germany, the US examinations appeared to be remarkably shallow. Were the European examinations really that much harder? If their curricula specifically addressed the topics included in the examinations, many student responses that appeared to require advanced analytical skills may, in fact, only have required recall of specific information treated in their coursework.

Non-US students, however, generally had substantially more class time devoted to the study of chemistry over more years of secondary schooling than did US students. Additionally, their instructional resources and opportunities for laboratory experiences were generally more abundant than in US schools. Consequently, chemistry examinations from other industrialized nations tended to be more challenging than those in the United States.

It would be dangerous and irresponsible for the United States (or any other country) to try to base its chemistry curricula and educational goals on those of another country that might appear to be "superior" as a result of this kind of analysis. On the other hand, the international educational community needs to be aware of chemistry curricula of other countries, since we have much to learn from each other.

5

Physics Examinations

Kjell Gisselberg

Overview of Examinations

THIS ANALYSIS OF PHYSICS matriculation examinations relied on 1991 and 1992 examinations from the countries and regions under consideration (see Table 1-1), with the following exceptions:

- Only the 1991 Israeli examination were included, as no translation of the 1992 examination was available.

- US examinations were from 1988 and 1993.

England and Wales (1991, 1992)

Advanced Level Examination, Associated Examination Board (AEB). This examination consisted of three papers and took 7 hours.

- Paper 1 took 2 hours and had 50 multiple-choice questions.

- Paper 2 took 3 hours and had four sections:
    - Section 1 had eight questions, all of which were compulsory.
    - Section 2 had one question.
    - Sections 3 and 4 both contained two questions, with students selecting one question from each section. Each of these questions described a situation or theme around which some 10 subquestions were asked.

- Paper 3 took 2 hours and consisted of two practical items, both of which were compulsory.

121

Advanced Level Examination, University of London (UOL) Examinations and Assessment Council. This examination consisted of four papers and took 8.25 hours to complete.

- Paper 1 was 1.5 hours long and had 45 multiple-choice questions.

- Paper 2 had two sections, which together took 2.5 hours:
  - Section 1 took 1 hour and had seven questions, all of which were compulsory.
  - Section 2 took 1.5 hours and had three questions, each of which had three parts. The first part of each question was compulsory; either the second or the third part also had to be answered.

- Paper 3 took 2 hours and had two sections:
  - Section 1 was a physics reading comprehension test, consisting of a one-page write-up on a specific topic—an application of a physical principle or phenomenon—and several questions related to the text.
  - Section 2 consisted of five questions; candidates had to answer any two.

- Paper 4 was a laboratory test, which took 2.25 hours. There were three versions of this paper, each with different but similar tasks. The paper was split into four parts, each of which took 30 minutes. The first three parts were three different experiments students were asked to perform; the fourth part was a data analysis question. The remaining 15 minutes were for writing up results.

Some boards included a classroom-based practical component in which a student's classroom project, assessed by his or her teacher, was included in the final A-level grade.

France (1991, 1992)

Baccalauréat Examination, Aix Region.
Baccalauréat Examination, Paris Region. The two French examinations were quite similar. In both, physics and chemistry items were presented as a combined examination that took 3.5 hours. The entire examination consisted of only seven questions, each of which had several subquestions; the physics portion comprised three of the seven questions. Throughout this chapter, the values of some French examination topics may seem inordinately high, but this resulted from the examinations being short and having few questions.

In the French examinations, problems often were given as "paper and pencil" experiments—that is, experiments were described, and students were asked to make the calculations connected with the investigations.

Germany (1991, 1992)

Abitur Examination, State of Baden-Württemberg. This examination took 4 hours and was comprised of four tasks, each of which addressed a separate topic. There were three or four questions associated with each topic.

Abitur Examination, State of Bavaria. This examination took 4 hours. It consisted of some 10 to 13 questions in each of six areas. The two German examinations were quite different from each other in terms of their content coverage. The content of the Bavarian examination seemed rather advanced. Many basic areas in physics were not covered at all, even if they were the basis for more complex areas that were covered. The Baden-Württemberg examination was more limited than the Bavarian one. Several major physics topics, such as heat, atomic physics, and nuclear physics, were not included on the Baden-Württemberg examination.

Israel (1991)

Bagrut Examination. This examination took 5 hours and consisted of three equally weighted papers given at three different occasions.

- Paper 1 took 1.5 hours and had nine different sections, each having two questions addressing a specific topic area, e.g., light and waves, modern physics, mechanics, relativity, and astrophysics. Students chose only two sections, and also chose between the two questions in each section.

- Paper 2 was a physics laboratory test, for which 2 hours were allotted. Students received credit for setting up the experimental system; conducting the experiment; and providing analysis, conclusions, and answers to specific questions.

- Paper 3 took 1.5 hours. It consisted of five questions on mechanics, three of which were compulsory.

Japan (1991, 1992)

Entrance Examination, Tokyo University. This examination took 2.5 hours and covered three major topic areas. With so few questions, it was difficult for the examination to obtain a wide coverage of the subject. In fact, its modern physics component consisted of only a few points on atomic physics and did not touch on either nuclear physics or the theory of relativity.

Typically, items on the Japanese examination began by describing a fairly complex situation and then asking a series of questions about it. Each of these situations thus covered more than one area of physics, thereby increasing the content area coverage beyond what would otherwise be possible with only a few main questions.

Sweden (1991, 1992)

National Examination. This examination took 3.75 hours and consisted of two sections.

* Section 1 had seven multiple-choice or short-answer questions.
* Section 2 had six extended-answer questions.

The Swedish National Examination has a different purpose than the examinations of other countries in this book, but it is still rather important both for the individual student and for the teachers. The results of the examination serve as an extended calibration for the level of the class, and provide a guideline for the teacher when he or she is going to give the students their final marks (which also include the teacher's own assessments). It is not necessary to test the students on the complete course, and the examination is given before atomic physics and nuclear physics have been covered. Nor are there any items in thermophysics.

United States (1988, 1993)

Students took either a Physics B or C examination, depending on whether they had taken an AP Physics B or C course.

Physics B Advanced Placement Examination. This 3-hour examination consisted of two equally weighted papers containing the topics in the AP Physics B course: mechanics or electricity and magnetism, kinetic theory and thermodynamics, waves and optics, and modern physics (College Board, 1994b). The mathematical demands of Physics B questions generally were limited to algebra and trigonometry.

• Paper 1 took 1.5 hours and comprised 70 multiple-choice questions.

• Paper 2 took 1.5 hours and had six free-response questions, each of which was divided into several subquestions.

Physics C Advanced Placement Examination. This 3-hour examination consisted of two papers. Students focused on either mechanics or electricity and magnetism, or both, depending on the topics in the AP Physics course they had taken (College Board, 1994b). Calculus was used to formulate physical principles and apply them to physical problems. Scores were reported separately for each of the two areas of emphasis.

• Paper 1 took 1.5 hours and was divided into two sections of 45 minutes each. Students, depending on their course emphasis, took one or both:
  - Section 1 consisted of 35 multiple-choice questions on mechanics.
  - Section 2 consisted of 35 multiple-choice questions on electricity and magnetism.

• Paper 2 took 1.5 hours and was divided into two sections of 45 minutes each. Students, depending on their course emphasis, took one or both:
  - Section 1 had three extended-answer questions, with several subquestions, on mechanics.
  - Section 2 had three extended-answer questions, with several subquestions, on electricity and magnetism.

Compared to all the others, the US Physics C examination covered mechanics and electricity in depth. It included advanced topics such as the dynamics of rigid bodies, Ampere's Law, and Gauss's Law—concepts covered in just a few of the other examinations.

General Structure

The following table summarizes key structural features of the various countries' examinations.

Table 5-1.   General Structure of the Physics Examinations
Number of Scorable Events (SEs)

Country/ Region	Exam Length (Hrs)	Total No. of SEs	Multiple-Choice (SEs)	Free-Response (SEs)	Laboratory Practical (SEs)
E&W - AEB	7.0	155	50	85	20
E&W - UOL	8.3	111	45	43	23
France*	2.1	11	—	11	—
Germany	4.0	38	—	38	—
Israel	5.0	129	—	96	33
Japan	2.5	25	—	25	—
Sweden	3.8	13	1	12	—
US	3.0	95	70	25	—

*   Physics questions comprise three out of five sections in combined chemistry/physics examinations.

Length and Scorable Events

The time required to sit for the physics examinations ranged from about 2 hours to more than 8, and the number of scorable events ranged from 11 (France) to well over 100 (both English and Welsh examinations and Israel).[1]

Choice

Most of the examinations offered students little choice: On the examinations from France, Germany, Japan, Sweden, and the United States, students were expected to answer all questions. On the other hand, parts of the examinations from England/Wales and Israel afforded students substantial choices among questions.

---

[1] Scorable events were the smallest, discrete questions in each examination. They were the unit of analysis in this study and are described in the Approach to Comparing Examinations section of Chapter 1.

The Israeli examination offered the most freedom, giving students choices in the two written papers, but not in the practical activity paper. In the first written paper, which accounted for one-third of the total examination score, there were nine sections, each of which contained two questions. Students only had to answer one question from each of two sections. In the third paper of the Israeli examination, students needed to answer three of five questions. The coverage of the Israeli examination was affected by its use of choice. Even though 33 content areas were covered in the examination's first paper, students only needed to answer two questions out of a possible eighteen. Thus, even if the examination covered many areas of physics, the emphasis accorded each of these areas was low because of the number of options. On the other hand, the third paper covered only mechanics. This topic accordingly accounted for a large proportion of the examination's total score. Not surprisingly, then, four of Israel's top five content areas were in the mechanics category; the fifth one, simple harmonic motion, was closely related to mechanics.

The England/Wales UOL examination offered the next-greatest degree of choice. About a quarter of the total examination score could be obtained in questions that were optional. Paper 2, section 2, had three 3-part questions. Students were to answer the first part of each of these questions and then *either* the second or the third part. In some cases, this entailed a choice between completely different content areas; e.g., choosing between questions about radioactivity or spring oscillations. In others, the choice was between different aspects of the same areas; e.g., explaining either sound or light-wave behavior. In paper 3, section 2, students could answer any two questions out of five. The five questions covered completely different content areas, e.g., energy and its uses, medical physics, etc.

The AEB examination in England/Wales also offered options; these made up 15 percent of the total examination score. Specifically, in both of the last two of four sections in paper 2, students chose one of two questions to answer. The choice here was between different content areas.

Item Characteristics

This section compares item types across examinations and discusses the use of diagrams, graphs, and tables in these items.

Item Types

The amount of each examination allocated to different item types varied greatly among the set of examinations. This important examination characteristic is described three ways. Table 5-1 already provided the number of scorable events found of each main item type. Similarly, Figure 5-1 shows the amount of each main item type but in terms of how much examination time was allotted to it.

Figure 5-1.   General Item Types in Physics Examinations

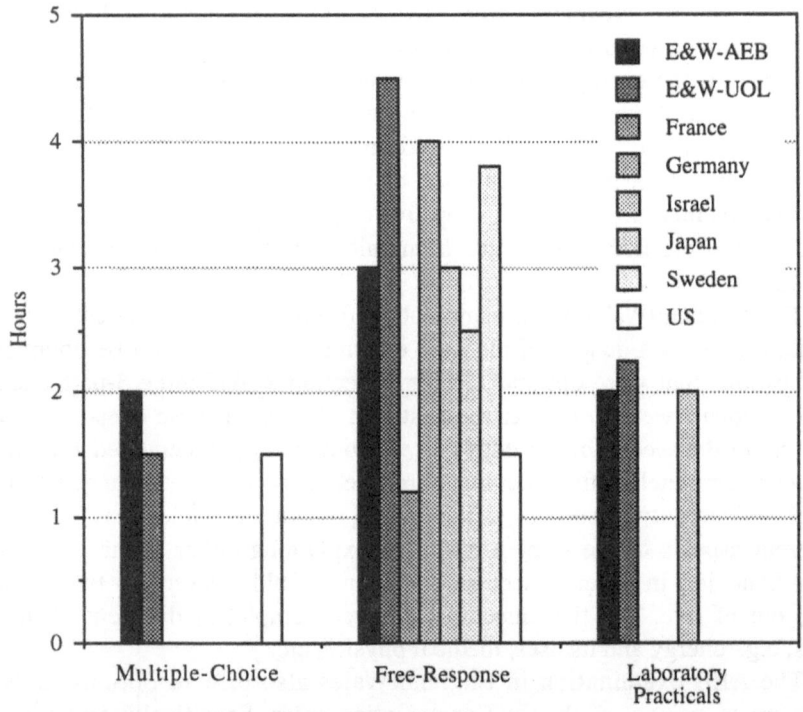

Finally, Table 5-2 shows the proportions of the total examination score accounted for by various item types, and it divides free-response items into short- and extended-answer categories.

Multiple-Choice Items. The United States, especially, and England/Wales used multiple-choice items. Other countries did not employ this item type at all.[2]

---

[2] However, since 1993, Sweden also has included a special multiple-choice section on its examination; previously, only a few multiple-choice questions could be found in the short-answer section of the examination.

Table 5-2.   Item Type in Physics Examinations
Percentage of Examination Score

Country/ Region	Multiple- Choice	Free- Response, Short- Answer	Free- Response, Extended- Answer	Laboratory Practical
E&W - AEB	17	38	14	31
E&W - UOL	14	12	50	23
France	—	33	66	—
Germany	—	19	81	—
Israel	—	24	42	34
Japan	—	39	61	—
Sweden	4	30	66	—
US	50	20	30	—

Free-Response Items. In the most common type of free-response item across all the examinations, students were asked to calculate the value of a specific physical quantity based on numbers provided for other physical quantities. In another common approach, however, questions represented given quantities by letters or symbols. Students had to derive an algebraic expression for a requested quantity, as illustrated by Example 5-1 from France. Students tend to find this type of item more difficult than when numbers are given.

Example 5-1.   (France, Aix Cluster, 1992)
Partial Question

An Instrument developed to study radioactivity uses $^{137}_{55}Cs$ as a radioactive source because it emits β and γ radiations.

b.   Using the definition of T and the decay law N(t), find, in function of T, the theoretical value of the decay constant γ. N(t) is the number of nuclei of a radioactive nuclide that are not disintegrated at the time t.

Question a precedes and question c follows.

The examinations varied greatly in terms of the number of items that required algebraic solutions. At the ends of the spectrum were the Japanese examination, on which algebraic solutions were required for about 80 percent of the total score, and the Swedish examination, which did not require algebraic solutions. Between these extremes were the US Physics C examination (algebraic expressions took 39 percent of the total score); the French, German, and Israeli examinations (algebraic expressions represented between 9 and 13 percent); and the English/Welsh and US Physics B examinations (algebraic expressions occupied only 5 percent or less).

Of the two US examinations, the Physics C examination contained the most items requiring algebraic solutions: This is in accordance with descriptions of this examination, which state that it requires a deeper mathematical sophistication than the Physics B examination. The French items did not seem to be as complex as the Japanese items. In the French items, students tended to be asked first to give an algebraic expression and then to calculate a numeric answer from given values of the different quantities.

In a few more cases, students were not presented with either numbers or algebraic quantities to be manipulated for an answer. Rather, they had to devise how to set up or solve a problem, as illustrated by Example 5-2 from England and Wales.

Essay Items. None of the examinations *required* an essay question. On the Israeli examination, however, students could choose to answer one or both of two essay questions. These questions, which took about 20 minutes each, were offered in two of the nine sections of paper 1.

Word-Phrase Items. Only one or two word-phrase questions were found in a few examinations. Because they constituted such a very small percentage of the total points available, the most reasonable portrayal of physics examinations is to report a virtual absence of these items.

Practical Activity Items. Hands-on experimental tasks were found in the examinations from Israel and the two examination boards in England/Wales. Although all three examinations devoted approximately the same amount of time to practical tasks, the maximum score given varied between one-third of the total examination score (Israel) to one-quarter or one-fifth (England/Wales).

In the AEB examination, students had 2 hours to complete two tasks that mainly consisted of using apparatus, taking measurements, and recording and graphing the data. The instructions for the work were very precise, and not much was left for the student to decide, as illustrated by the partial labo-

Example 5-2.   (England and Wales, Associated Examining Board, 1991)
Partial Question

This question is about the design of experiments to measure the speed of an air-gun pellet. The speed of the pellet is known to be about 40 m s^{-1} and the mass of the pellet is about 0.5 g.

One student suggests that momentum ideas might be used. It is proposed that the pellet be fired into a trolley of mass M and that the speed v of the trolley after impact be determined by finding the time it takes a card to cross the path of a light beam. The light beam illuminates a photodiode which controls a timer. The timer can record the time that the light beam is cut off to the nearest 0.01 s. The system is shown in Fig. 1.

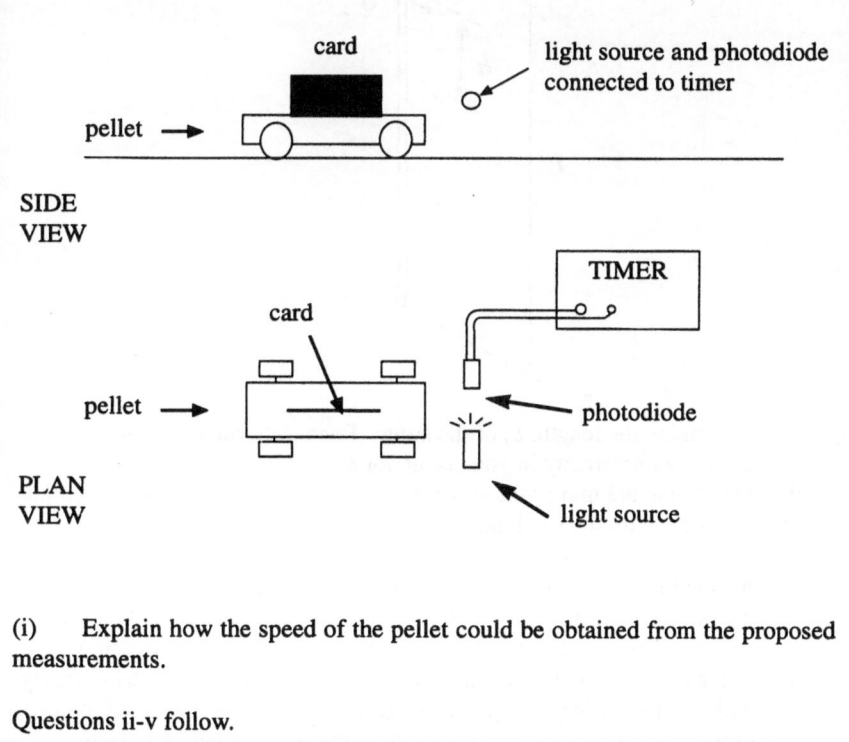

(i)   Explain how the speed of the pellet could be obtained from the proposed measurements.

Questions ii-v follow.

ratory practical in Example 5-3. The nature of the laboratory practicals can be gleaned from the points allocated in the scoring guide: Recorded observations, 18; Appropriate repetition of observations, 5; Graph of observations, 7; Deductions from results, graph, or data, 11; and General presentation and consistency, 4.

Example 5-3.   **(England and Wales, Associated Examining Board, 1992)**
                    Partial Question

1. You are to investigate the oscillations of a sand-filled straw as it swings in a
    vertical plane about a horizontal axis. You will be asked to obtain two val-
    ues for the acceleration of free fall, g.

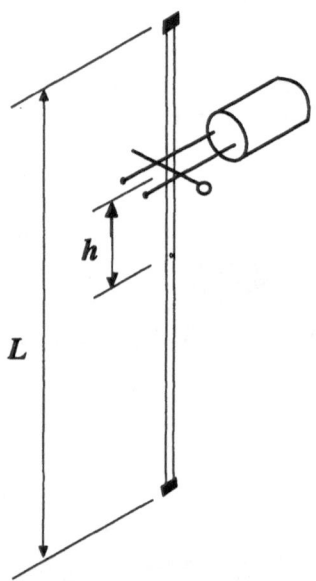

a.   i.   Determine the length, L, of the straw. Record its value in mm.
      ii. State the uncertainty in your value for l.
      iii. Determine the mid-point of the straw **carefully** and mark it by piercing
           the straw with the small pin.

b.   i.   At a point approximately 30 mm from the mid-point of the straw, push
           the pin through the straw at right angles to it. Measure and record, in
           mm, the distance, h, between the pin and the mid-point of the straw.
      ii. Arrange the straw as shown in the diagram so that it can swing freely
           with the pin acting as a pivot. Determine, T, the period of oscillation
           of the straw as it swings, with small amplitude, in a vertical plane.

[Question 1 continues with parts c-g that require students to tabulate and graph
data obtained in parts a-b, above, and to perform calculations using the data.]

In the UOL examination, three similar versions of the practical activity were assigned to students, presumably at random. Each version was 2.25 hours and consisted of three experiments and one data analysis question; All four parts received equal weight. Some tasks in the UOL examination seemed less prescriptive than those in the AEB examination, and more explanations of actions taken were requested, as illustrated by the partial laboratory practical in Example 5-4.

Example 5-4.   (England and Wales, University of London)
Partial Question

You are to make some basic measurements on a glass tube in order to determine the density of the glass from which it is made. You do not have access to a balance. You should aim for as much precision in your measurements as possible. (italics in original)

Take measurements to establish the length, *l*, and external diameter, *d*, of the glass tube as precisely as you can. You are provided with a metre rule and vernier callipers, and a micrometer screw gauge is available. Record all your measurements and the instruments you use below.

Hence calculate the external volume, $\pi d^2/4\ l$ of the tube.

Estimate the uncertainty in each of your measurements. State, with reasons, which of these uncertainties will have the greater effect on the uncertainty of your value for the volume.

To find the internal volume of the tube, you are provided with some water, a small measuring cylinder and a rubber bulb. Fix the rubber bulb on the end of the glass tube and use it to suck some water into the tube. The water can then be transferred to the measuring cylinder. Hence determine the internal volume of the tube. Record your measurements below.

[Students execute more procedures and then determine density of the glass.]

The 2-hour Israeli practical task consisted of three parts. In part 1, the students set up the experimental system (for a maximum of 10 points); in part 2, they conducted the experiment (40 points); and in part 3, they were asked to graph and analyze their data, draw conclusions, and answer questions (50 points). Although the task descriptions in parts 1 and 2 were very detailed, part 3 was much "looser" and students had to explain their conclusions.

Use of Diagrams, Graphs, and Tables

Diagrams, graphs, and tables were widely used on physics matriculation examinations. (Figure 5-2.) The Japanese examination made the most use of these devices, with all scorable events having either an accompanying diagram or table. The German and Israeli examinations made the least use of these devices.

Figure 5-2.   Use of Diagrams, Graphs and Tables in Physics Examinations

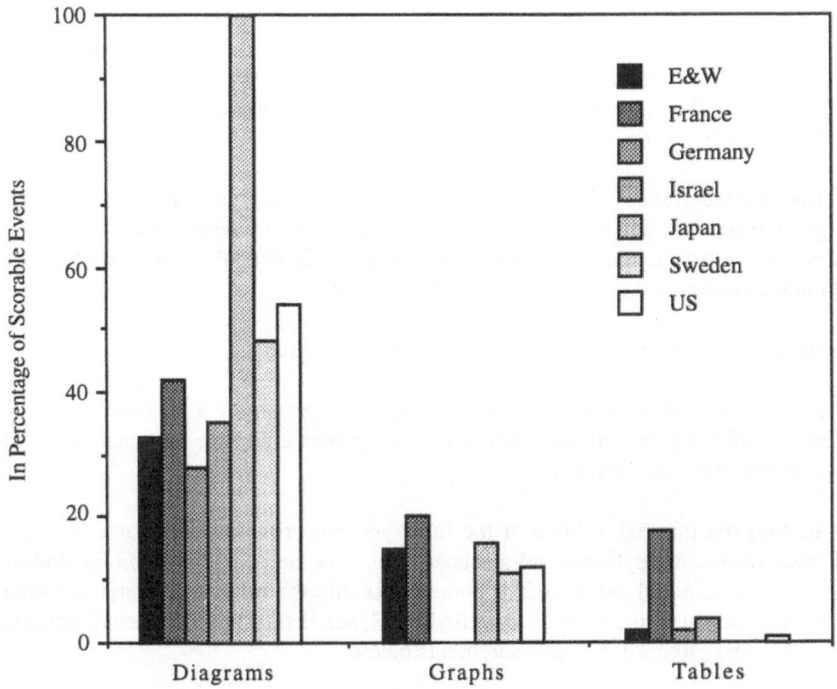

Diagrams. Ordinary diagrams were by far the most common type of illustrative material, although there was great variation in their use between countries and examinations. Diagrams were often used to clarify a situation so that the students could understand it. Sometimes they played a crucial role in the presentation of a problem or a situation.

In the Japanese examination, all items involved diagrams. At the other extreme was the UOL examination, for which diagrams illustrated only 18 percent of the items.

Graphs and Tables. In general, graphs were more widely used than tables. In fact, in all countries except France, tables were hardly used at all. On the other hand, graphs were used only to a very limited extent in the German and Israeli examinations. Tables and graphs were generally used for one purpose: the students collected information from them and used it in their work with the corresponding question.

Examination Topics

This section compares the examinations' topics in two complementary ways. The first subsection notes contrasts only among the most emphasized topics of each examination while the second subsection indicates all physics topics covered by each examination. The first comparison, dealing with each examination's five most emphasized topics, looks at more detailed topics such as capacitors, potential voltage and resistance, or electronics and semiconductors. The comprehensive reporting of all topics across all examinations begins with a focus on more general topics, such as mechanics or electricity, but then continues by comparing whether examinations addressed the more detailed topics associated with these general topics.[3]

Emphasized, Detailed Topics in Each Examination

Overview of Most Emphasized Topics. Table 5-3 shows the top five content areas—as determined by percentage of examination score—of each examination. These areas are listed in descending order of frequency.

Although the top five topics listed in Table 5-3 are quite specific, it is nevertheless interesting to discover that none of them were common to all examinations. In fact, only basic topics in electricity (electric field, potential, voltage, resistance; charges in electric and magnetic fields), harmonic motion, and optics (reflection and refraction) occurred in four or five examinations. Simple harmonic motion accounted for a particularly high percentage of the Israeli examination's coverage. This strong emphasis is attributable to the scoring weight of the country's practical activity: investigating the motion of a physical pendulum comprised one-third of the total examination score. Optics was particularly dominant in the Tokyo University exami-

---

[3] Appendix A may be helpful for understanding the meaning of the more general topics because it lists the specific topics associated with them. General topics were taken from the Science Framework of the Third International Mathematics and Science Study (Robitaille et al., 1993). The detailed topics were added to permit more specific descriptions of the examinations.

Table 5-3.   Physics Topics, Detailed: Five Most Emphasized in Each Examination
Percentage of Examination Score

	AEB	UOL	Aix	Paris	Bav	B-W	Isr	Jap	Swe	US-B	US-C
Electric Field, Potential, Voltage, Resistance	10	4	5	11						9	13
Simple Harmonic Motion	11		8			7	30	6			
Reflection and Refraction			15					20	14	11	
Charges in Electric and Magnetic Fields					5	9		7		5	
DC Circuits	7								12		6
AC Circuits			36	11		22					
Electromagnetism				14					8		7
Linear Motion		3					8	12			
Laws of Circular Motion				10			8	9			
Heat and Energy		4		15							
Electromagnetic Spectrum, Diffraction		5				13					
Angular Momentum, Moment of Inertia								11			8
Work, Energy, Power									6		
Kinetic & Potential Energy									10		

Table 5-3.   Continued

	AEB	UOL	Aix	Paris	Bav	B-W	Isr	Jap	Swe	US-B	US-C
Superposition of Waves					**11**						
Infrasonic and Ultrasonic Waves			4								
Charging & Discharging of Capacitors	8										
Induction										7	
K-M View of Gases						6					
Photoelectric Effect						6					
Line Spectra						8					
First Law of Thermodynamics										7	
Kinetics of Nuclear Decay				**26**							
Static Equilibrium	8										
Motion in Two Dimensions								4			
Laws of Linear Motion										7	
Minkowsky Space					**11**						
*Total-Topic Concentration	43	21	90	61	35	62	60	54	50	40	40

Bold indicates values of 10 percent or greater. Blank cells indicate only that the topic was not one of the top five in the examination. The examination still may have included the topic. Topics listed in decreasing order of Topic Frequency; i.e., the number of examinations having the topic.

* Topic Concentration indicates the percentage of the examination comprised by its five most emphasized topics.

nation. Both the Aix regional cluster examinations and those from Baden-Württemberg strongly emphasized questions involving AC circuits.

As seen from Table 5-3, more than half of the top five topics were on only one examination's list. Among these less frequent topics, the ones constituting the greatest percentages of an examination were kinetics of nuclear decay (26 percent, Aix regional cluster), superposition of waves (11 percent, Baden-Württemberg), and Minkowsky space (11 percent, Bavaria).

Topic Concentration. The summary row of Table 5-3 roughly indicates the examinations' topic concentration. It shows the percentages of the various examinations comprised by their five most emphasized topics.[4] The most focused examinations were in the Aix regional cluster (90 percent concentration). In England and Wales, the United States, and the Bavarian state, only 21-40 percent of the examinations' topics were accounted for by the five most emphasized topics; conversely, the majority of these examinations were devoted to a lighter treatment of many other topics. The topic concentration of all other examinations was distributed within the range of 50-62 percent.

All Topics Covered by Each Examination

General Topics. Table 5-4 illustrates the overall topic orientation of the physics examinations analyzed. These orientations are determined by the extent to which the examinations featured items from classical physics (mechanics; electricity; magnetism; waves, sound, and light; and thermophysics) and modern physics (atomic and quantum physics, nuclear physics, and relativity and cosmology).

Three examinations—the US Physics B, Bavarian, and UOL examinations—had items in all areas, thereby covering the physics field comprehensively. The AEB and Israeli examinations included almost all topics, and were consequently listed as comprehensive examinations. The AEB examination did not address relativity and cosmology; the Israeli examination did not address magnetism.

The other cluster in Table 5-4 is comprised of those examinations that concentrated exclusively, or mostly, on classical physics topics. The Tokyo University examination was essentially classical; it included all five classical physics topics and only gave slight attention to one area of modern physics. The Baden-Württemberg, US Physics C, and Paris examinations covered

---

[4] This statistic can be thought of as an indicator of an examination's breadth in topic coverage. Examinations with higher and lower topic concentrations focused on fewer or more topics, respectively.

Table 5-4.   Physics Topics, General: All Topics Found in Each Examination
Percentage of Examination Score

Country/ Exam	Classical Physics					Modern Physics		
	Mech- anics	Elec- tricity	Magne- tism	Waves	Thermo- physics	Atomic, Quantum Physics	Nuclear Physics	Relativity, Cosmol.
Comprehensive Topic Coverage								
E&W - AEB	35	32	4	13	6	3	4	—
E&W - UOL	27	17	6	23	14	7	4	1
Ger. - Bav.	17	2	7	7	14	23	12	19
Israel	42	4	—	36	6	6	2	5
US - B	29	19	9	19	10	9	3	2
Mostly Classical Physics								
Fr. - Paris	29	27	29	15	—	—	—	—
Ger. - B-W	15	32	12	40	—	—	—	—
Japan	41	10	7	33	6	2	—	—
Sweden	40	14	17	26	—	—	—	3
US - C	49	28	21	2	—	—	—	—
Fr. - Aix	5	40	12	24	—	—	29	—
Topic Average	30	20	10	22	5	4	5	3

four of the five classical physics areas. (The absent topic was thermo-physics.) The Swedish examination also mainly covered classical physics, but included some items on the Theory of Relativity as well. Because this examination is given about four months before the end of the physics course, it does not reflect the full course content. Specifically, atomic and quantum physics as well as nuclear physics are taught after the examination has been administered. Those countries in Table 5-4 that did not include modern physics topics in physics examinations also did not include them in their chemistry examinations (see nuclear chemistry and quantum theory in Table 4-4).

The Aix examinations did not fit into either the comprehensive or classical clusters. Of the classical topics, they did not include thermophysics, and weakly treated magnetism. They gave considerable attention to nuclear physics, but other modern physics topics were not addressed.

The remainder of this section details examination coverage for each of the eight general physics topics of Table 5-4. The reported topics are based on classifying items into a total of 105 subtopics.

Mechanics. Mechanics was the most common topic in the examinations surveyed, making up an average of 30 percent of the score in all the examinations. Table 5-5 shows the distribution of subtopics within mechanics. Physical properties and fluid behavior were little represented on the examinations. In fact, the former area was only addressed in a practical activity on the UOL examination, where students were asked to determine the density of a material. Apparently, this area was generally considered too basic to test. Similarly, physics examinations did not tend to address fluid behavior, since this category covered both concepts learned much earlier (e.g., Archimedes' Principle) and advanced topics (e.g., the Continuity Equation and the Bernoulli Theorem), which are often not taught below the university level.

The remaining four topics within mechanics were covered in all but the Aix examination. The AEB examination emphasized types of forces rather than time, space, and motion and the dynamics of motion; the reverse was true for the Israeli and Japanese examinations, and also—to a certain extent—for the Paris examination. Within energy types, sources, and conversions, items about kinetic and potential energy were found in all but the Paris examination. Items in this area about work, energy, and power and about energy types and energy transformations were mainly represented in the two US and two English/Welsh examinations. In the types of forces topic, items on nuclear forces—a fairly advanced topic—were found only in the US Physics B and Bavarian examinations. Static equilibrium items were covered in the US, English/Welsh, and Paris examinations.

The US and English/Welsh examinations did not seem to cover circular motion—a topic within the dynamics of motion content area—as much as did the German, Israeli, Japanese, and Swedish examinations. Also in this area, the US Physics C examination and the Israeli examination both put great emphasis on angular momentum, moment of inertia, and rotational kinetic energy. (See Example 5-5.) These topics are fairly advanced and did not appear in the other examinations save for a few items in the AEB examination.

Table 5-5.   **Coverage of Mechanics Topics**
Percentage of Examination Score

Country/ Exam	Energy Types, Sources, Conversions	Types of Forces	Time, Space, & Motion	Dynamics of Motion	Fluid Behavior	Physical Properties	Total
E&W - AEB	1-5	**18**	1-5	6-10	1-5	—	35
E&W - UOL	1-5	6-10	6-10	1-5	1-5	1-5	27
Fr. - Aix	1-5	1-5	—	—	—	—	1-5
Fr. - Paris	1-5	1-5	**14**	6-10	—	—	29
Ger. - Bav.	1-5	1-5	1-5	6-10	1-5	—	17
Ger. - B-W	1-5	1-5	1-5	6-10	—	—	15
Israel	1-5	1-5	**15**	**23**	1-5	—	42
Japan	6-10	1-5	**17**	**14**	—	—	41
Sweden	**16**	6-10	1-5	**13**	—	—	40
US- B	6-10	6-10	6-10	**11**	—	—	29
US- C	6-10	**12**	6-10	**19**	—	—	49

Bold indicates values greater than 10 percent. Table values 1-5 and 6-10 indicate that average values for two years of these examinations fell within these ranges. Values less than 0.5 percent are not included.

Example 5-5.   **(United States, Physics C, 1988)**
Full Question

A figure skater is spinning on frictionless ice with her arms fully extended horizontally. She then drops her arms to her sides. Which of the following correctly describes her rotational kinetic energy and angular momentum as her arms fall?

	Rotational Kinetic Energy	Angular Momentum
A.	Remains constant	Remains contant
B.	Decreases	Increases
C.	Decreases	Decreases
D.	Increases	Decreases
E.	Increases	Remains constant

Electricity. Electricity was a key category in most examinations, accounting for an average of about 20 percent of the score on the physics examinations. Table 5-6 reports the coverage of electricity topics. In the Aix examination, this topic accounted for 40 percent of the total points. At the other extreme, however, were the Bavarian and Israeli examinations. Only 1.6 percent of the Bavarian examination's total points covered electricity; moreover, since these mainly dealt with charges in electric and magnetic fields, they might be said to relate more to magnetism than to electricity. On the Israeli examination, electricity accounted for less than 4 percent of the maximum score.

In general, the US, English/Welsh, Japanese, and Swedish examinations emphasized DC circuits more and AC circuits less. In the Japanese and US examinations, there were, in fact, no AC circuit items at all. All of the items on the Aix examination dealt with AC circuits. This topic also figured strongly both on the Baden-Württemberg examination and the Israeli examination, where it was essentially the only stressed topic in the electricity category. Some argue that this emphasis reflects reality, since our electric power supply almost always operates with alternating currents. Others hold that AC circuits and electronics are not pure physics topics, but rather technical applications of basic electricity and magnetism.

The English/Welsh and US examinations touched on several other aspects of the electricity category, as did the Baden-Württemberg examination. This latter examination was the only one to address electromagnetic oscillations, as illustrated in Example 5-6.

Example 5-6.   **(Germany, Baden-Württemberg, 1991)**
Partial Question

a.   A capacitator is charged by means of a switch in position 1 across a resistance R and then, in position 2, discharged across a coil. Draw the circuit diagram for this trial. Explain why, in position 2, there is electromagnetic oscillation. Why is the oscillation damped? Now an undamped oscillation is to be produced by reverse coupling. Draw a circuit diagram.

b.   Write a differential equation of the undamped electromagnetic oscillation.

The rest of question b and questions c and d follow.

Table 5-6. Coverage of Electricity Topics
Percentage of Examination Score

Country/Exam	Charge, Conductors, Insulators	Potential, Voltage, Resistance	Capacitors, Dielectrics	Charging, Discharging Capacitors	Power and Energy	DC Circuits	AC Circuits	Electronics, Semi-conductors	Electro-magnetic Oscillations
E&W - AEB	1-5	6-10	—	6-10	1-5	6-10	1-5	6-10	—
E&W - UOL	1-5	1-5	1-5	1-5	1-5	1-5	1-5	1-5	—
Fr. - Aix	—	1-5	—	—	—	—	**36**	—	—
Fr. - Paris	—	**11**	—	1-5	—	1-5	11	1-5	—
Ger. - Bav.	1-5	1-5	—	—	—	1-5	1-5	—	—
Ger. - B-W	—	1-5	1-5	1-5	1-5	1-5	**22**	—	1-5
Israel	—	—	—	—	—	1-5	1-5	—	—
Japan	—	1-5	—	—	—	6-10	1-5	—	—
Sweden	—	—	—	—	—	**12**	1-5	—	—
US - B	1-5	6-10	1-5	—	1-5	1-5	1-5	1-5	1-5
US - C	1-5	**13**	1-5	1-5	1-5	6-10	—	1-5	—

Bold indicates values greater than 10 percent. Table values 1-5 and 6-10 indicate that average values for two years of these examinations fell within these ranges. Values less than 0.5 percent are not included.

Magnetism. Magnetism played a key role in the Paris examination, although it had a much lower presence in others, resulting in an overall average of about 10 percent of total examination scores. (See Table 5-7). The US Physics C, English/Welsh, and Swedish examinations had items in four or five of magnetism's topics, while the Israeli examination had no items at all on this subject.[5] The Paris examination put a decided emphasis on the first three content topics—magnetic forces and magnetic fields, electromagnetism, and induction. In contrast, the German examinations had few or no items in these areas, but instead emphasized charges in electric and magnetic fields. All the examinations, except for the Israeli and the French examinations, included items in this topic; questions dealt mainly with charged particles in mass spectrometers or similar phenomena. The German, Japanese, and US examinations allocated more than 5 percent to such topics.

Table 5-7.   **Coverage of Magnetism Topics**
Percentage of Examination Score

Country/ Exam	Magnetic Forces, Magnetic Fields	Electro-magnetism	Induction	Self-inductance	Charges in Electric & Magnetic Fields
E&W - AEB	—	—	1-5	1-5	1-5
E&W - UOL	1-5	1-5	—	—	—
Fr. - Aix	—	—	—	1-5	—
Fr. - Paris	6-10	**14**	6-10	—	—
Ger. - Bav.	—	—	—	—	6-10
Ger. - B-W	—	—	1-5	1-5	6-10
Israel	—	—	—	—	—
Japan	—	—	—	—	6-10
Sweden	—	6-10	1-5	6-10	1-5
US - B	—	1-5	1-5	—	6-10
US - C	1-5	6-10	6-10	1-5	6-10

Bold indicates values greater than 10 percent. Table values 1-5 and 6-10 indicate that average values for two years of these examinations fell within these ranges. Values less than 0.5 percent are not included.

---

[5] Values for some topics in England/Wales were less than 0.5 percent and are not reported in the table. The text refers to these values because even such small percentages of these very lengthy examinations can represent one or two scorable events.

Waves, Sound, and Light.[6] The percentage of the examination score ranged from over 40 percent in the Baden-Württemberg exam down to almost zero in the US-C examination, with an average of more than 20 percent. Because the US-C course covers only mechanics and electricity, no attention to this topic was expected. The second lowest coverage occurred in the German examination from Bavaria, where this topic comprised about 7 percent of the total score. In Table 5-8, a few subtopics have been omitted because their percentage of the examination score was very low and, they occurred in only one or two examinations. Infrasonic and ultrasonic waves were in the UOL examination, but just as a reading comprehension question about ultrasonic cleaning. Intensity of sound in the Israeli examination, radio transmission in the UOL examination, fibre optics in the UOL and Japanese examinations, and polarized light in the UOL examination also had very low percentages—one percent or less—and were consequently omitted.

The subtopics that carried the most points were the following: simple harmonic motion; diffraction and the electromagnetic spectrum; and reflection and refraction. Harmonic motion is covered on Israeli and AEB examinations by their practical tasks—the Israeli examination contains an investigation of a pendulum, and the AEB practical addresses spring oscillations. Diffraction was extensively tested in the examinations from Baden-Württemberg, AEB, and Sweden. Ray optics was more common in the Japanese, French, US-B, and Swedish examinations. In the Paris examination, all items in this category covered reflection and refraction. Example 5-7 is a rather involved optics item from the Japanese examination.

Thermophysics. Thermophysics was a low emphasis on many of the examinations, making up an average of 5 percent in total examination scores. In fact, Table 5-9 indicates that five examinations—the Baden-Württemberg, Aix, Paris, Swedish, and US Physics C examination—did not have any items at all on this topic. This lack may be because this subject was covered on these countries' chemistry examinations. Of those countries that included thermophysics on their physics examinations, England and Wales—in both their examinations—covered all five content areas. Both the AEB and UOL examinations gave the most points for heat and temperature items.

---

[6] Many of the subtopics within sound and light often are applications of more general wave phenomena. Interference in thin layers and diffraction are often connected with light, but diffraction also occurs with water waves. Standing waves in strings and pipes are often treated within sound but, in fact, they are applications of the reflection and superposition of waves. Thus, no effort has been made to split this topic into separate categories for waves, sound or light.

Table 5-8.   Coverage of Waves, Sound and Light Topics
Percentage of Examination Score

Country/ Exam	Simple Harmonic Motion	Transverse Waves	Longitu- dinal Waves	Superposition of waves, interference	Doppler effect	Diffraction, electomag- netic spectrum	Standing waves	Reflection, refraction	Light intensity, luminosity
E&W - AEB	**11**	—	—	—	—	1-5	1-5	—	—
E&W - UOL	1-5	1-5	1-5	1-5	—	1-5	—	1-5	1-5
Fr. - Aix	6-10	—	—	—	—	—	—	**15**	—
Fr. - Paris	—	—	—	—	—	—	—	**15**	—
Ger. - Bav.	—	—	—	—	1-5	1-5	—	1-5	—
Ger. - B-W	6-10	6-10	—	**11**	1-5	**13**	—	1-5	—
Israel	**30**	—	—	—	—	1-5	—	1-5	1-5
Japan	6-10	—	1-5	—	1-5	—	—	**20**	—
Sweden	1-5	1-5	1-5	1-5	—	6-10	—	**14**	—
US - B	1-5	1-5	—	1-5	1-5	1-5	1-5	**11**	—
US - C	1-5	—	—	—	—	—	—	—	—

Bold indicates values greater than 10 percent. Table values 1-5 and 6-10 indicate that average values for two years of these examinations fell within these ranges. Values less than 0.5 percent are not included.

Example 5-7.   (Japan, 1992)
                      Partial Question

The diagram represents the cross section ABCD of a clear glass column, which is supported to maintain horizontally the surface AD.

The left portion of the cross section is 1/4 of a circle with origin at 0. AD and BC are parallel to each other. Surfaces BC and CD are made out of frosted glass.

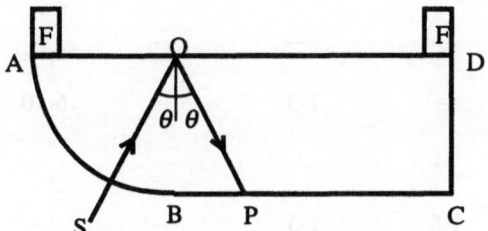

A ray of light was aimed at point 0 from a point s on the surface AC. The incident ray made an angle of incidence, 0, but the reflected light ended up hitting point P on surface BC. Let the glass refractive index be N_o.

Answer the questions below: Note that in this question, you must assume that the refractive index is a relative refractive index with regards to air. When angle θ was increased gradually from 0°, suddenly at angle "θ_o", the brightness of P intensified.

Explain this phenomena and describe the relationship between N_o and θ_o.

Next, an edge F was made. When the same experiment was performed as above, with an incident ray, which made an angle of $\theta < \theta_o$, then 2 points appeared on the surface BC.

Name the point closer to B, P_1, and the other P_2.

Their intensity of brightness is stronger than any other time. In the following questions, just consider these two points. What kind of reflection caused P_1? and P_2? When the angle of incidence, θ, was gradually increased from 0° to a point where $\theta = \theta_o$, one of the points started to shine brighter.

Which point was it?

Questions 4-6 follow.

Table 5-9.   Coverage of Thermophysics Topics
Percentage of Examination Score

Country/ Exam	Heat and Temperature	Physical Changes	Explanation of Physical Changes	Kinetic Theory	Energy & Chemical Change
E&W - AEB	1-5	1-5	—	1-5	1-5
E&W - UOL	6-10	1-5	—	1-5	1-5
Fr. - Aix	—	—	—	—	—
Fr. - Paris	—	—	—	—	—
Ger. - Bav.	—	1-5	—	6-10	1-5
Ger. - B-W	—	—	—	—	—
Israel	1-5	1-5	—	—	1-5
Japan	—	1-5	—	—	1-5
Sweden	—	—	—	—	—
US - B	1-5	—	—	—	6-10
US - C	—	—	—	—	—

Table values 1-5 and 6-10 indicate that average values for two years of these examinations fell within these ranges. Values less than 0.5 percent are not included.

The Bavarian and US Physics B examinations covered four of the five thermophysics topics.

Atomic, Nuclear, and Quantum Physics. Atomic and quantum physics were not well-represented on these examinations. (See Table 5-4.) Information that was covered primarily applied to quantum theory, although there were also a few items on chemical changes; atoms, ions, and molecules; and subatomic particles. The Bavarian and US Physics B examinations emphasized quantum physics more than any other countries. Most of the quantum physics items in these examinations were on photoelectric effect or line spectra.

Nuclear physics was unevenly handled across the examinations. The UOL and Bavarian examinations covered most of the topics associated with nuclear physics. Also, more than a quarter of the total score for the Aix examination was for nuclear physics. (This high percentage, however, is due to the fact that there were only three main questions on the French examination, one of which—in both years for which examinations were analyzed—dealt with a nuclear physics issue.) On the other hand, there were no nuclear physics items on the Paris, Israeli, Japanese, Swedish, or US Physics C examinations. On those examinations that covered nuclear physics, the most common topics were alpha and beta particles, neutron/proton ratios, and kinetics of nuclear decay.

The UOL and Israeli examinations were the only ones that covered biological effects of radiation. Example 5-8 was a sample UOL question.

Example 5-8.   **(England and Wales, University of London, 1992)**
Partial Question
Question i precedes

ii. As well as exposure to natural background radiation, a person living in the United Kingdom may receive doses of γ-radiation from medical treatment. Describe, stating their purposes, *two different medical procedures* involving γ-photons. (Do **not** give details of detection or imaging devices which are used in your chosen treatments.)

Relativity and Cosmology. Other physics topics that were under-represented in the examinations were relativity and cosmology. (See Table 5-4.) Cosmology was only covered in a few items on the Israeli examination, including the one on the big bang theory provided in Example 5-9. The theory of relativity was covered somewhat more frequently across the examinations, although the majority did not address the area.

Example 5-9.   **(Israel, 1992)**
Full Question

a. What is the theory of the "big bang"? List two observational findings which support the theory of the big bang.

b. How is it possible to estimate the time, T, which has passed since the big bang? Explain.

c. It is commonly argued that the universe whose average density is less than a certain density called "critical density P_c" will expand forever. Explain this argument.

d. Develop a formula which will make it possible to estimate the critical density P_c of the universe. (Base your calculations on conservation of mechanical energy and the Hubble constant H.)

The Bavarian examination heavily emphasized relativity, particularly in such advanced topics as Minkowsky space,[7] which accounted for almost 11 percent of the total examination score. (See Example 5-10.) This examination also covered relativistic energy and momentum. This topic, which accounted for about 5 percent of the total Bavarian score, was accorded some items on the Swedish examination as well.

Applications and Other Secondary Topics. The primary topics of all the physics examinations' questions were subjects within the field of physics. A very few, however, had secondary topics that would not normally be regarded as physics subjects. The TIMSS framework describes these topics as follows:

- Applications of science in mathematics, technology
- History of science and technology
- Nature of scientific knowledge
- Science and other disciplines

A mere 17 scorable events addressed these topics, almost all of them being *applications of science in mathematics and technology*. Although there was one question in the US Physics B examination and one in the AEB examination, the items were mainly found in the University of London examination. (See Example 5-11.)

Varied Treatments of Questions. The ways that examinations structured and phrased questions can be gleaned from the examples given thus far where one or two examples were provided to illustrate a topic or other examination feature. To further illustrate the examinations' varied questioning strategies, the following set of examples on the same topic is provided. Except for the French and Israeli examinations, all examinations included questions on charged particles in electric and magnetic fields. In addition to sharing the same topic, these questions required similar performance expectations. (See next section.) Several required Understanding Simple Information, e.g., asking for the direction of a force acting on a charged particle in an electric field. Many questions also expected students to Apply Scientific Principles to Solve Quantitative Problems, e.g., calculating masses, velocities, distances, etc.

---

[7] Minkowsky was a physics professor at the University of Güttingen, which is near— although not in—Bavaria. The item on the Bavarian examination covering Minkowsky space may be less advanced than it seems. For one thing, the students may be prepared to expect the question; also, it may be possible to solve it without a deep understanding of it.

Example 5-10.   (Germany, Bavaria, 1992)
Full Question

1.   In a fictitious universe, the speed of light in a vacuum is c = 15.0 m/s. Otherwise, all laws of physics still hold, particularly those of the theory of relativity. A standing observer Be stands along a long, straight street, which defines a frame of reference S (x ; t); Be stands at x = 0. A streetlight is located 90 m from Be in the positive x direction. A streetcar travels towards the streetlight at a constant speed v = 9.0 m/s and passes Be. The driver Fa sits at the front of the streetcar exactly at the zero point of the frame of reference S' (x' ; t') of the streetcar. At that moment when Fa passes Be (event V), the synchronized clocks of the S-system and the S'-system are set to zero ($t_v = t_v' = 0$).

a)   Sketch a perpendicular coordinate system t-x:

t-axis: 1 s $\overset{\Delta}{=}$ 1 cm ; -4 s $\leq$ t $\leq$ 8 s

x-axis: 15 m $\overset{\Delta}{=}$ 1 cm ; -45 m $\leq$ x $\leq$ 135 m

Sketch the world line of the driver Fa and calibrate it as the time axis of the S' system. Also, sketch and calibrate the x'-axis of the S' system.

b)   At time $t_R = 2.0$ s the streetlight turns "red" (event R). Sketch R into the coordinate system and calculate the coordinates of R in the S'system. Compare the spatial relationship and the temporal behavior of V and R in both systems.

c)   Sketch into the coordinate system the S' point in time $\overset{\wedge}{t_R}{}'$ at which the driver Fa sees the event R.
Calculate $\overset{\wedge}{t_R}{}'$.

The streetlight emits "red" light having a frequency of 4.2 (10^{14}) Hz. Even in this fictitious universe, the frequencies ranging from 4.0 (10^{14}) Hz to 4.7 (10^{14}) Hz are considered "red".

d)   Using calculations, determine whether the driver Fa can recognize the "red light" as "red".

e)   A speed limit is to be instituted in front of streetlights so that upon approach the light frequency deviates no more than 15% from the emitted light frequency. Calculate the speed limit necessary to meet this requirement.

f)   In the system S' the streetcar is 45 m long. What is the length of the streetcar to the observer Be?

g)   Clocks are located in the middle of the streetcar and in the end of the streetcar which also can be read by Be. Event M occurs when the clock in the middle of the streetcar passes Be, and event E occurs when the clock in the end of the streetcar passes Be. Calculate the time coordinates of the events M and E in both systems.

Example 5-11.   (England and Wales, University of London, 1991)
Partial Question
Question a and the beginning of question b precede.

b. Which of the following stress/strain curves can be associated with which of these groups (all are drawn to the same scale)?

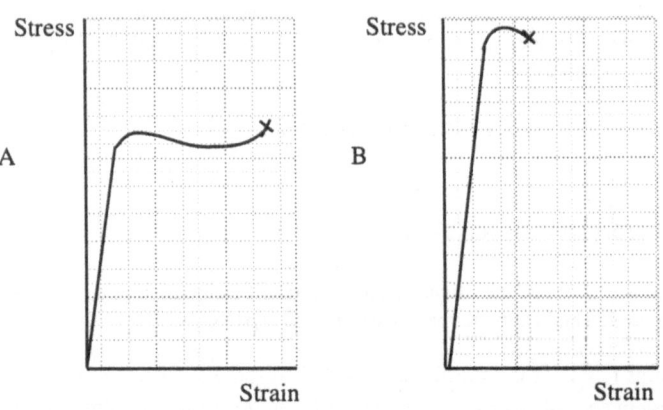

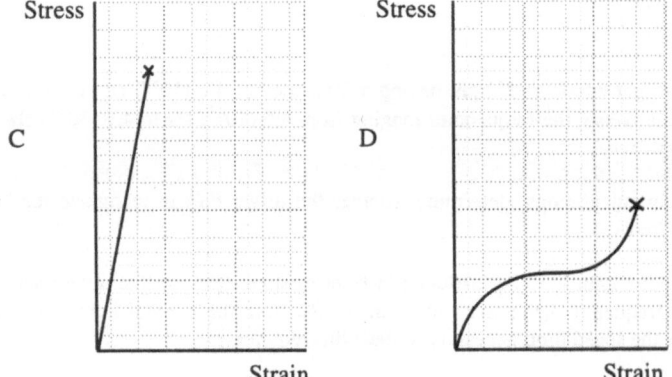

Identify the material of each of the following with one of the above stress/strain curves: glass window, pvc floor tiles, inflatable rubber boat, nylon fishing line.

c. Microscopic examination of work-hardened polycrystalline metals reveals grain boundaries. Explain what is meant by *work-hardened*.

The common context in these questions was charged particles being acceler-ated in an electric field, the use of a speed filter consisting of crossed electric and magnetic fields, and circular motion in a magnetic field. Questions on five examinations were similar and represented the simplest treatment of the topic—the two English/Welsh examinations (AEB, UOL), the two US exam-inations (US-B, US-C), and Baden-Württemberg. Example 5-12 illustrates that these questions only addressed fundamental aspects of the topic. They consisted of several scorable events, most of which were quite narrow and highly structured. Only one law or one or two formulas were needed to answer them. However, the last scorable event in Example 5-12 and that of some other examinations listed above required a somewhat deeper under-standing. Further, the Baden-Württemberg question also introduced particle motion in a direction inclined to the magnetic field which caused the particle to move in a spiral path.

Example 5-13 from Bavaria had fewer but more complex scorable events. There was no force acting in the direction of the motion and thus no increase in speed. First the charged particle was moving in a magnetic field and then in an electric field. Both fields were oriented so that the path of the particle was part of a circle. Further, relativistic effects had to be calculated, which adds to the complexity of the question.

Example 5-14 from Japan was a completely different type of question. The charged particle was suspended by a string and moved as a conical pen-dulum in a vertical magnetic field. Thus a component of the string tension and the magnetic force acting on the charged particle combined to form the centripetal force. In further questions, the students were asked to calculate the angular frequency as a function of the magnetic field and to conclude what would happen when the magnetic field increased. This context was very complex and solving the problem required a very good understanding of both physics and mathematics.

The seven questions mentioned above were all scaffolded—a sequence of scorable events where each question built upon previous ones. Scaffolding can be used to assess a topic in a more complete way and to give credit for partial knowledge. Example 5-15 from Sweden was a single question, but an involved one in which the context was only generally specified. Students had to assume that the particle had a certain kinetic energy, use this expression together with the formulas for circular motion in a magnetic field, and find out what quantity they needed to measure and how to use it. There was no guidance provided by a sequence of subquestions—just the information that the radii of the circles could be measured in the diagram.

Example 5-12.   (US, AP-C, 1993)
Full Question

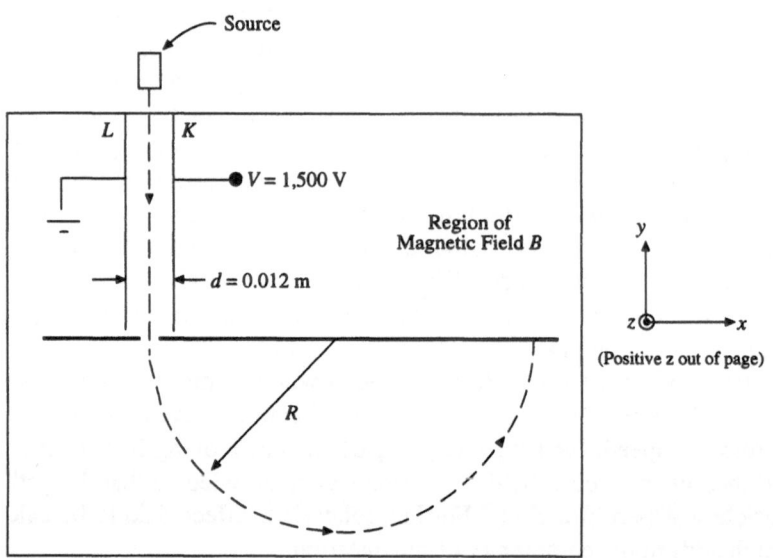

3.   A mass spectrometer, constructed as shown in the diagram above, is to be used for
     determining the mass of singly ionized, positively charged ions. There is a uniform
     magnetic field $B = 0.20$ tesla perpendicular to the page in the region indicated in the
     diagram. A potential difference $V = 1,500$ volts is applied across the parallel plates
     L and K, which are separated by a distance $d = 0.012$ meter and which act as a
     velocity selector.

   (a)   In which direction, relative to the coordinate system shown above on the right,
         should the magnetic field point in order for positive ions to move along the
         path shown by the dashed line in the diagram above?

   (b)   Should plate K have a positive or negative voltage polarity with respect to plate
         L?

   (c)   Calculate the magnitude of the electric field between the plates.

   (d)   Calculate the speed of a particle that can pass between the parallel plates with-
         out being deflected.

   (e)   Calculate the mass of a hypothetical singly charged ion that travels in a semi-
         circle of radius $R = 0.50$ meter.

   (f)   A doubly ionized positive ion of the same mass and velocity as the singly
         charged ion enters the mass spectrometer. What is the radius of its path?

Example 5-13.   (Germany, Bavaria, 1992)
          Full Question

1. The mass m and the velocity v of electrons is to be experimentally determined using the apparatus sketched below:

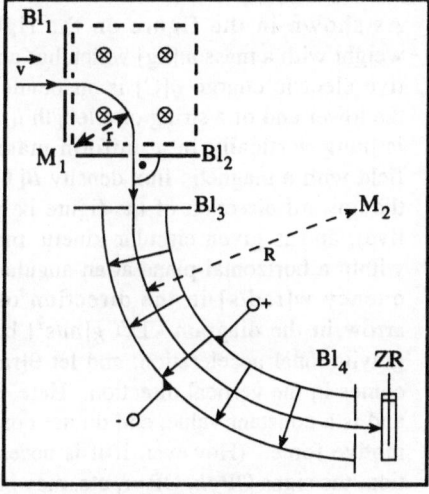

   The electrons first pass through a uniform magnetic field having the flux density B along an arc shaped path with radius r. After passing through the filter Bl_2 and Bl_3, the electrons enter a capacitor enclosed with cylindrical plates in which the electric field strength E is set so that the electrons travel along a second arc with radius R.

   The electrons emerging from the filter Bl_4 are recorded by a counting tube ZR.

   (a) Explain why the electrons pass through the entire apparatus at a constant velocity.

   (b) Formulate the general equation for the velocity v and for the mass m of electrons, relative to r, R, B and E. [partial answer: v = E R / r B]

   (c) Calculate the velocity v and the mass m of the electrons for

   r = 0.500 m, R = 2.00 m, B = 7.84 mT and E = 5.39 (10⁵) V/m

   Show that the results prove the relativistic relationship between m and v, and calculate the kinetic energy of the observed electrons in keV.

Example 5-14.   (Japan, 1991)
Full Question

1. As shown in the figure on the right, a
    weight with a mass m[kg] which has a posi-
    tive electric charge q[C] is suspended on
    the lower end of a string of a length l[m]. It
    is hung vertically in a uniform magnetic
    field with a magnetic flux density B[T] (let
    the upward direction of the figure be posi-
    tive), and is given circular kinetic motion
    within a horizontal plane at an angular fre-
    quency w[rad/s] in the direction of the
    arrow in the diagram. Let g[m/s^2] be the

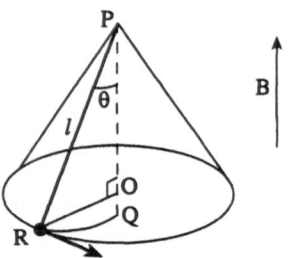

    gravitational acceleration, and let θ[rad] be the angle to which the string
    comes in the vertical direction. Here, assume that θ does not depend on B
    and is a constant value, and do not consider the effect of dielectric electro-
    motive force. (However, if it is necessary in the explanation of the solu-
    tion, the signs OP=h, OR=r, etc. may be used.)

I.. (1) Find the formula which indicates the balance of forces related to the
         horizontal direction of the weight (OR direction). Also, explain the
         directions of the respective forces and their meanings, when, in partic-
         ular, the magnetic field is in the upward direction (B>0).
    (2) From the results to the previous question (1), calculate the angular fre-
         quency w by a function of B.
    (3) Investigate what kind of value the angular frequency w will approach
         in relation to sufficiently positive and negative values of B, and in a
         graph express w as a function of B (positive and negative).
II. (1) If the magnetic field is sufficiently large in the upward direction, as a
         rule, what force(s) will come into equilibrium with what other
         force(s)? Explain.
    (2) If the magnetic field is sufficiently large in the downward direction,
         how would it be then?
    (3) If the magnetic field is sufficiently large in the upward direction, inves-
         tigate which is larger, the positional energy of the weight U or the
         kinetic energy K. However, let the standard for the positional energy
         be the value at the minimum position Q of the weight when the string
         is plumb.
    (4) Consider the same even when the magnetic field is sufficiently large in
         the downward direction.

Example 5-15.   (Sweden, 1991)
Full Question

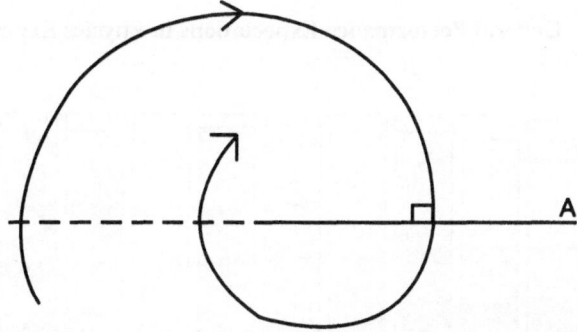

A charged particle moves in vacuum in a plane perpendicular to a uniform magnetic field. It strikes a thin lead foil A and passes through it. What percentage of its kinetic energy does the particule lose in this process? Measurements may be made in the diagram which is drawn in some (unknown) scale. The charge of the particle is constant and its velocity is much less than the speed of light.

Performance Expectations

The physics examinations stressed students' skills in theoretical problem solving. Thus, more than half (62 percent) of all available points on each examination were allocated to theorizing, analyzing, and problem solving, as shown in Figure 5-3. An additional 21 percent of all points were awarded for understanding information. The US and English/Welsh examinations gave more points to understanding information than did the other examinations.

Regarding the scope of performance expectations, the Swedish and US examinations assessed performance in only three of the four categories; no items on these tests addressed Investigating the Natural World. All the other examinations addressed all four performance expectation categories.[7]

England and Wales put a lesser priority on Theorizing, Analyzing, and Solving Problems than did the other countries. Instead, England and Wales, together with the United States, awarded more points for Understanding information. They primarily accomplished this through multiple-choice items. About 60 percent of the multiple-choice items on examinations fell

---

[7] Performance Expectations, categories used in the Third International Mathematics and Science Study to describe the student behaviors elicited by various types of questions, are further described in the text accompanying Tables 1-3 and 2-2.

into the Understanding category. The Swedish and Japanese examinations most heavily emphasized the Theorizing, Analyzing, and Problem-Solving category.

Figure 5-3.   General Performance Expectations in Physics Examinations

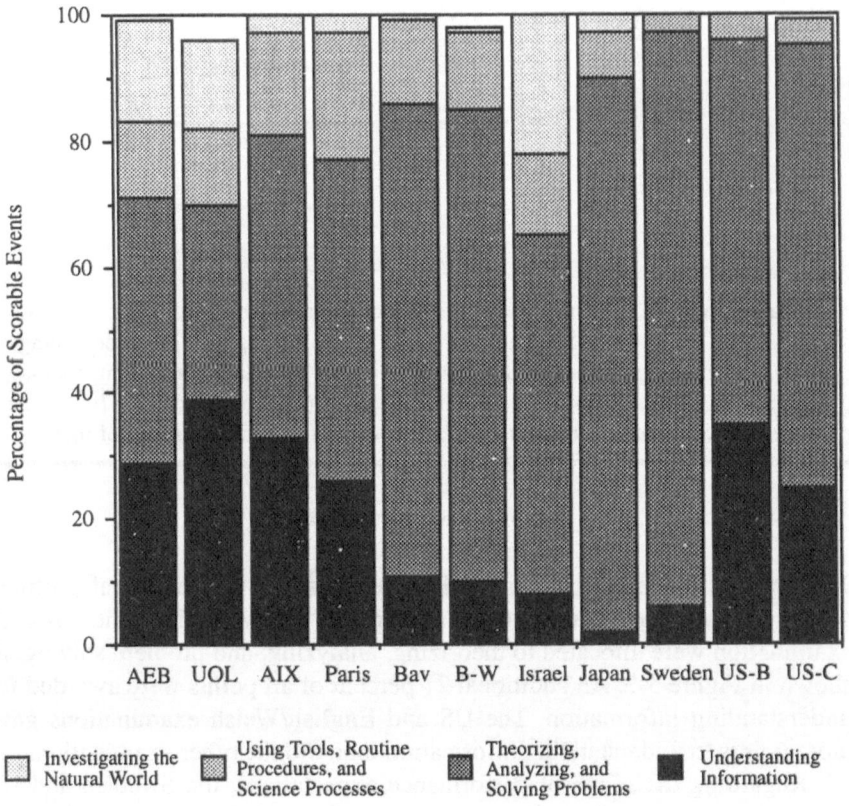

A special type of performance was tested in the UOL examination, which featured a set of physics reading comprehension items. (See Example 5-16.) Students read a physics-related passage, and then were asked a series of questions—almost all of the answers to which could be found in the text passage. Students needed to understand what they had read; they also needed some knowledge of certain physics concepts.

Example 5-16.   (England and Wales, University of London, 1991)
Partial Question

Metals and metallic alloys contain electrons, some of which are free to move at random (at speeds of about $10^6 ms^{-1}$) within a lattice of ions. The density of these free electrons is of the order of $10^{28} m^{-3}$ for a good conductor. If one end of a metal bar is heated, the electrons (and the ions) at that end gain energy. The electrons inevitably move away from that end, and their places are taken by slower electrons. This is simply a diffusion process taking place in a thermal gradient.

A different mechanism allows non-metals to conduct. This can involve only the atoms, since in non-metals there are no free electrons. At the heated end of a non-metallic bar, the atoms gain vibrational energy, and this is passed on to the neighboring atoms by elastic waves which, like sound waves, travel at the speed in the material (about 5000 ms^{-1}), but are of a much higher frequency (typically 10^{13} Hz). The atoms can absorb and emit energy only in quanta of magnitude *hf*. By analogy with quanta of light (photons), these quanta are called phonons. The energy carried by the phonons does not reach the far end as fast as their high speed would suggest, since they are readily absorbed by the atoms and re-emitted (they are said to be scattered) because they naturally have frequencies similar to those of the vibrating atoms. So the phonons diffuse through the material, and the randomness of this process means that they can take a long time to cover a short distance. Electrons, in metals, are weakly scattered and thus their diffusion rate is much higher. Hence metals are generally better thermal conductors than non-metals.

This background information continues in three additional paragraphs that are used for items 3-7.

1.   (a) Two types of energy quanta are mentioned in the passage; what are they called?
     (b) Give an example of a physical quantity other than energy which is quantized.
2.   What is meant by (a) a thermal gradient, and (b) the scattering of phonons?

Questions 3-7 follow.

Understanding, and Theorizing, Analyzing, and Solving Problems

Table 5-10 details the Understanding, and Theorizing, Analyzing, and Solving Problems performance categories.

Table 5-10.   Detailed Performance Expectations in Physics Examinations: Understanding Information, Analyzing and Solving Problems
Percentage of Examination Score

Country	Understanding Information		Analyzing and Solving Problems	
	Simple Information	Complex Information	Quantitative Problems	Developing Explanations
E&W	4	29	28	6
France	2	27	44	5
Germany	—	10	62	14
Israel	—	8	45	11
Japan	—	2	77	11
Sweden	—	6	91	—
US	1	29	65	—

Most of the examinations put a heavy emphasis on solving problems. In the Swedish examination, 91 percent of the total score required quantitative problem-solving skills. The English/Welsh examinations put the least emphasis on quantitative problem solving.

Using Tools, Routine Procedures, and Science Processes

Activities in this category were straightforward, essentially routine scientific procedures; these included:

- using apparatus, equipment, and computers;
- conducting routine experimental operations;
- gathering data;
- organizing and representing data; and
- interpreting data.

The English/Welsh and Israeli examinations were the only ones that included hands-on activities in their testing. In all three examinations, students had to use apparatus, perform routine experimental operations, and gather data.

Investigating the Natural World

Tasks that comprise Investigating the Natural World are less routine and require more creative thinking than those addressed in the preceding set of performance expectations. Specifically, these performance expectations include the following:

- designing investigations,
- conducting investigations,
- interpreting investigational data, and
- formulating conclusions from investigational data.

Various practical and written activities can address the first subcategory. The last three subcategories refer either to practical activities, including interpretation of and conclusion from these activities, or to the description of less routine experimental activities.

There was obviously an interest in and desire to stress experimental work in the examinations, but practical difficulties can prevent this from occurring. Three examinations required hands-on activities. In two of these, the students were also asked to interpret investigational data; in one, they were asked to formulate conclusions. In several examinations, students were asked to design experiments. The French examinations also asked students to interpret data and formulate conclusions. As stated before, neither the US nor the Swedish examinations had any items covering this performance expectation category.

Summary

School physics, as evidenced by these examinations, was mainly classical physics. Mechanics, electricity, magnetism, wave motion (waves, sound, and light), and thermophysics accounted for most of the total score in almost all of the examinations. In fact, in all but two examinations, the percentage devoted to these topics was over 80 percent; in three examinations, it was 100 percent.

Of the five classical subject areas, mechanics, electricity, magnetism, and wave motion were generally the most popular. Three examinations had items in these four areas only. There were some significant differences, however. Electricity was essentially excluded from both the Bavarian and Israeli examinations, and there was little or no coverage of magnetism in either the Israeli or Aix examinations. In contrast, the Paris examination had a very high percentage of its total score in the magnetism category.

Modern physics—atomic and quantum physics, nuclear physics, and relativity and cosmology—was fully represented in four examinations, and partly represented in another four. The amount of modern physics included on the examinations varied, and, in most cases, was low. In nine examinations, modern physics items accounted for less than 15 percent of the total score. The Bavarian examination was one exception: more than half of its total score was in the area of modern physics. In the Aix examination, the proportion of modern physics items was about 30 percent, based primarily on the strong emphasis on nuclear decay.

There was great variation in the topics covered in the examinations. A few examinations included more rarely seen topics, such as the Bernoulli Theorem, the Minkowsky diagrams, and the kinetic gas theory. Some included questions from all or almost all of the eight main areas of physics, classical and modern. The examinations that require more time can accommodate such broad coverage. Other examinations are more limited; these tend to concentrate on fewer main areas in the classical physics area. There are various reasons that certain areas are not included in certain examinations. For example, in the US Physics C examination, it is clearly stated that the course leading to this examination covers just two main topics, mechanics and electricity. In Sweden, atomic and nuclear physics are taught after the physics examination is taken, so these areas are not covered on the examination.

The examinations also differed considerably with regard to item type. In the US examinations, about 40 percent of the total scores were accounted for by multiple-choice items. All of the other examinations—except the English/Welsh—had very few or no such items. The English/Welsh examinations did offer quite a few multiple-choice items, but since these examinations took much longer and included many more items than the US examinations, the percentage of total score contributed by multiple-choice items was relatively small.

Almost all of the physics examinations contained items that refer to experimental situations. However, the problems were seldom contextually connected to an experimental investigation. Students were sometimes asked to design experiments or draw conclusions from fictitious data, but not very often. Two countries featured practical, hands-on activities. These tasks were strictly delineated and allowed the student very little freedom. The scoring was also very prescribed. One examination paper set forth the exact number of points awarded for making measurements, for repeating the measurements a sufficient number of times, etc. Such prescriptiveness is of course necessary for quick and reliable scoring. However, preselected apparatus, careful instructions of what to measure and how, and ready-made table shells and graphs do not represent actual experimental conditions. The importance of an experimental approach is emphasized by physics educators the world over because physics is an experimentally based and empirically founded science, whose laws are based upon experiments.

The performance expectations emphasized did not vary much among countries. All of the examinations primarily emphasized theorizing, analyzing, and solving problems. In Japan and Sweden, around 90 percent of the examinations' total scores were accounted for by items in this category. Most of these items involved applying scientific principles to solve quantitative problems. Obviously, this strong emphasis on problem solving leaves little room for practical activities or for questions in the understanding information category.

One examination—the UOL examination—had a slightly different distribution of scores over the performance expectation categories. A large portion of its scores were for practical tasks, which were—in comparison with those of other examinations—somewhat more open and less guided. The UOL examination also had more questions than the other examinations in the understanding category and fewer in the problem-solving category. Finally, the examination featured a unique set of questions that tested students' ability to read and understand a text with physical or technical content.

Overall, most of the examinations were strongly geared toward problem solving. The stress on classical physics is perhaps linked to this strong emphasis on problem solving. It is easier to construct a problem-solving item in mechanics or electricity than in atomic or nuclear physics. Thus, the tradition of classical physics and the tradition of physics as a problem-solving endeavor combine to yield a particular examination structure.

6

Mathematics Examinations

John Dossey

Overview of Examinations

THIS ANALYSIS OF MATHEMATICS matriculation examinations relied on the 1991 and 1992 examinations from the countries and regions under consideration (see Table 1-1), with the following exceptions:

- No translated Israeli mathematics examinations were available for study.

- US examinations were from 1988 and 1993.

England and Wales (1991, 1992)

Although different in construction, the two England/Wales examinations were quite similar in content and format. The examinations made use of the fact that students had also studied physics as part of their secondary schooling. They tested the connections between mathematics and physics in applied situations, especially in the items assessing students' knowledge of elementary analysis.

Advanced Level Examination, Associated Examination Board (AEB). From the eight mathematics papers that AEB offers, students typically take the set of two papers that correspond to the particular courses of mathematics they have studied. The following set of "Pure and Applied Mathematics" papers corresponds to the most popular of several mathematics programs. The other programs are Applied, Pure, Statistics, Pure and Statistics, Applied and Statistics.

This examination consisted of two papers and took 6 hours.

- Paper 1 took 3 hours and was comprised of two sections. Students answered all eight items in the first section and three of four items in the second. The paper covered the topic areas of functions and differential and integral calculus.

- Paper 2 also took 3 hours and consisted of two sections. Students answered all seven items in the first part and four of six in the second. In this paper, students were asked to apply the content measured in the first paper to physics-related situations.

- In both papers, students could use a calculator and metric graph paper in constructing their responses to the various items.

Advanced Level Examination, University of London (UOL) Examinations and Assessment Council. From the eight mathematics papers that UOL offers, students typically take the set of three papers that correspond to the particular courses of mathematics they have studied. The following set of papers correspond to "Mathematics," the most popular of five mathematics programs. The other four programs are Further Mathematics, Pure, Pure with Statistics, and Applied.This examination consisted of three papers and students were allotted 6.25 hours.

- Paper 1 took 2.5 hours and consisted of 15 compulsory items dealing with the application of differential and integral calculus.

- Paper 2 took 1.25 hours and consisted of 30 multiple-choice items covering functions and relations, differential and integral calculus, enumeration, and recursion.

- Paper 3 took 2.5 hours and contained eight items, of which students had to answer six. The paper's problems involved applications of calculus and probability and statistics.

Students were provided with a booklet, *Mathematical Formulae Including Statistical Formulae and Tables*, to use in completing the examination items.

France (1991, 1992)

Baccalauréat Examination, Aix Region. The examination of this regional cluster of academies consisted of three sections and took 4 hours.

All questions were compulsory.

-   Section 1 had three items covering complex numbers, similitudes in the complex plane, and other aspects of geometric transformations.
-   Section 2 consisted of two items dealing with parametric curves and conic sections.
-   Section 3 consisted of seven items covering elementary functions, limits, derivatives and their application, and sequences.

Baccalauréat Examination, Paris Region. This examination consisted of three sections and took 4 hours. All questions were compulsory.

-   Section 1 had two items that covered conic sections/complex numbers.
-   Section 2 had two questions that covered plane geometry viewed from a linear algebra perspective.
-   Section 3 had ten questions that covered elementary functions/calculus/geometric transformations.

Germany (1991, 1992)

Abitur Examination, State of Baden-Württemberg. This examination comprised nine sections—which took 4 hours. All questions were compulsory.

-   Sections 1-3 covered differentiation and integration and contained four, one, and four items respectively.
-   Sections 4-6 covered geometric transformations and contained four, three, and four items respectively.
-   Section 7 contained three dice-related probability problems.
-   Section 8 contained four applied conditional probability problems.
-   Section 9 contained four applied quality-control-related problems.

Abitur Examination, State of Bavaria. This examination took 4 hours and consisted of six sections—two each on differential and integral calculus, probability and statistics, and analytical geometry. All questions were compulsory.

-   Sections 1-2 covered differentiation and integration and contained two and five items respectively.
-   Sections 3-4 covered probability and statistics and contained three items each.

- Sections 5-6 covered analytical geometry and contained two items each.

Japan (1991, 1992)

Entrance Examination, Tokyo University. This examination took 2.5 hours and consisted of six questions, all of which were compulsory. The examination covered the areas of algebra and geometry, concepts and applications of functions, differential and integral calculus, and basic probability and statistics. Given the small number of questions, the coverage in a topic area was very narrow, but deep. Typically questions on the Japanese examination described a complex situation and then asked one or more questions about it. The increased depth provided by this structure separated students by their command of the various topic areas.

Sweden (1991, 1992)

National Examination. This examination took 3.75 hours and consisted of two sections.

- Section 1 had five items, each of which was compulsory.
- Section 2 had five items, each of which was compulsory.

The Swedish National Examination covered a broad range of content, but not at the depth covered on the examinations of other countries. In part, this was due to the purpose of the Swedish examination. The results of this examination are used as a guideline for schools about the level of their classes and to teachers as a guideline for the assignment of final marks. As such, the examination is not as high stakes as the examinations in other countries.

United States (1988, 1993)

Students take either a Calculus AB or Calculus BC examination, depending on the nature of the calculus course they have completed and their estimation of their level of achievement. The former examination covers the content of a one-semester university-level course in calculus, while the latter examination covers a full-year, two-semester sequence of university-level calculus.

AB Calculus Advanced Placement Examination. This 3-hour examination consisted of two papers, and covered material in the AP

Calculus AB course: the theory of functions, differential calculus, and elementary integration (College Board, 1994a).

- Paper 1 consisted of 45 multiple-choice questions to be answered in 1.5 hours.

- Paper 2 contained six free-response questions to be answered in 1.5 hours.

BC Calculus Advanced Placement Examination. This 3-hour examination consisted of two papers and covered materials in the AP Calculus BC class: the theory of functions, differential calculus, integration, and series of constants (College Board, 1994a).

- Paper 1 had 45 multiple-choice questions and took 1.5 hours.

- Paper 2 had six free-response questions to be answered in 1.5 hours.

General Structure

The following table summarizes key structural features of the various countries' examinations.

Table 6-1.   General Structure of Mathematics Examinations
Number of Scorable Events (SEs)

Country/ Region	Exam Length (Hrs)	Total No. of SEs	Multiple- Choice (SEs)	Free- Response (SEs)	Practical (SEs)
E&W - AEB	6.0	65	—	65	—
E&W - UOL	6.3	120	30	90	—
France	4.0	29	—	29	—
Germany	4.0	52	—	52	—
Japan	2.5	10	—	10	—
Sweden	3.8	12	—	12	—
US	3.0	64	45	19	—

Length and Scorable Events

The amount of time allotted to the mathematics examinations ranged from 2 1/2 hours in Japan to about 6 hours in England and Wales, with the others falling between 3 and 4 hours. Although, outside of England and Wales, all the countries had a fairly similar time span for the examinations, the number of scorable events varied greatly.[1] For example, France, Japan, and Sweden had only 29, 10, and 12 scorable events respectively, while Germany and the United States had 52 and 64 scorable events respectively.

Choice

Students had to work all the problems on the examinations from France, Germany, Japan, Sweden, and the United States; they were given no choice or options. In marked contrast, students had a great deal of choice on the two English/Welsh examinations. In the AEB examination, almost two-thirds of the total possible points were associated with item sets in which students could choose from among several problems. In one set, worth 30 percent of the total points, students chose three out of five problems to work; in the other, worth 33 percent, they selected two out of three problems. The UOL examination also gave choices on more than two-fifths of its possible points. This choice came in a paper in which students had to answer six out of eight questions.

Item Characteristics

This section compares item types across examinations and discusses the use of diagrams, graphs and tables in these items.

Item Types

The amount of each examination comprised by different item types varied greatly from one examination to another. This important examination characteristic is described three ways. First, Table 6-1 provided the number of scorable events found of each main item type. Second, Figure 6-1 also shows the amount of each main item type but in terms of how much examination time was allotted to it.

---

[1] Scorable events were the smallest, discrete questions in each examination. They were the unit of analysis in this study and are described in the Approach to Comparing Examinations section of Chapter 1.

Figure 6-1.   General Item Types of Mathematics Examinations

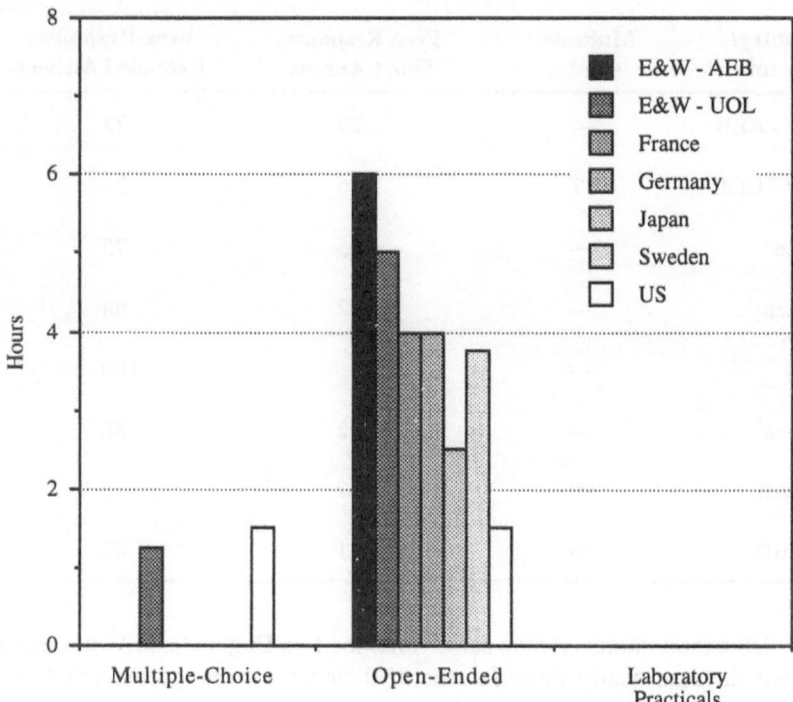

Third, Table 6-2 shows the percentage of the total examination score accounted for by various types of items: multiple-choice, and both short- and extended-answer, free-response. Perhaps the most striking difference in item type usage is that only three of the examinations used multiple-choice questions-the AEB and the two US Advanced Placement examinations. The amount of multiple-choice items on the US Advanced Placement examinations was directed toward increasing topic coverage in the time allotted, and allows equating of scores across years.

Another interesting difference was the balance on each examination of short- and extended-answer items among the free-response items. Note the difference in the percentage of points in the two US examinations between short- and extended-answer, free-esponse items. This reflects the inherent view among American mathematics educators that the BC examination is a more demanding and rigorous examination. While the UOL examination contained some multiple-choice items, the percentage of final score they affected was far less than the 50 percent controlled by multiple-choice items

Table 6-2.   Item Type in Mathematics Examinations
Percentage of Examination Score

Country/ Exam	Multiple- Choice	Free-Response, Short-Answer	Free-Response, Extended Answer
E&W - AEB	—	29	72
E&W - UOL	17	6	78
France	—	25	75
Germany	—	12	88
Japan	—	—	100
Sweden	—	12	88
US - AB	50	22	28
US - BC	50	1	49

in the US examinations. At the same time, the two England and Wales examinations devoted nearly three-fourths of their total scores to extended free-response items. Across all examinations, the Japanese (100 percent), Germans (88 percent) and Swedes (88 percent) devoted the greater percentages of their scores to extended free-response items, while the United States devoted the least with BC calculus (49 percent) and AB calculus (28 percent).

Multiple-Choice Items. Only the UOL and US examinations made use of the multiple-choice format. On the US examinations, there was a single correct response for each question. The response format called for students to select this response from among a set of five possible responses, as did some UOL items. Other UOL items, however, had a more complex response format, requiring students to select all of the responses that applied.

Examples 6-1 and 6-2 are two items dealing with the derivative and differential equations. The first item, taken from a US-BC calculus examination reflected a rather straight-forward multiple-choice item. The second item, taken from a University of London examination provided a multiple-choice setting where students could select one or more of the possible responses as being correct through use of the combinations shown in the answer scheme.

answer scheme. Note that the latter item also forced students to consider more possibilities as part of the item.

Example 6-1.   (US, AP, Calculus BC, 1993)
                    Full Question

39. If $\dfrac{dy}{dx} = \dfrac{1}{x}$, then the average rate of change of y with respect to x on the closed interval [1, 4] is

(A) $-\dfrac{1}{4}$ (B) $\dfrac{1}{2}\ln 2$ (C) $\dfrac{2}{3}\ln 2$ (D) $\dfrac{2}{5}$ (E) 2

(Correct Answer: C)

Example 6-2.   (England and Wales, University of London, 1992)
                    Full Question

Directions: For each of the following questions, ONE or MORE of the responses given are correct. Decide which of the responses is (are) correct. Then choose

A if **1, 2,** and **3** are correct
B if only **1** and **2** are correct
C if only **2** and **3** are correct
D if only **1** is correct
E if only **3** is correct

Directions Summarized			
A	1	2	3
B	1	2	
C		2	3
D	1		
E			3

29. $f(x) \equiv 2xe^{-x}$.

   1 $f(x)$ has a minimum value at $x = 1$.

   2 $y = f(x)$ satisfies $e^{x}\left(y + \dfrac{dy}{dx}\right) = 2$.

   3 $y = f(x)$ satisfies $\dfrac{d^2y}{dx^2} + 2\dfrac{dy}{dx} + y = 0$.

(Correct Answer: C)

It is interesting that in England and Wales, the AEB examinations did not have multiple-choice items, while the UOL examinations did. On the US examinations, the multiple-choice items were weighed equally with the free-response items to determine the final-scale score for the examinations. Hence, while multiple-choice items constituted 70 percent of the examinations' scorable events (from Table 6-1), they only used 50 percent of the response time and were allocated 50 percent of the total examination score.

Free-Response Items. Most of the examinations relied heavily on extended-answer, free-response items. Moreover, answers to the majority of these items depended on answers given to other free-response items. This "building" effect indicates the in-depth focus on particular topics that many of the examinations used. Example 6-3 illustrates an extended free-response question with a "building" structure. While part c seemed to depend on part b, a student could have answered it directly via the relationship such a graph must have with the line y = x. However, many students might work toward c from their answer to part b, possibly risking the loss of additional points depending on the grading scheme applied.

Example 6-3.   **(England and Wales, University of London, 1992)**
Full Question

4. The function f is given by $f : x \longrightarrow e^{-2x}$, $x \, \varepsilon R$.

(a) State the range of f.

(b) Define the inverse function f^{1} in a similar manner to f.

(c) Sketch the curve with equation $y = f^{1}(x)$.

It is not clear from the scoring rubrics what various examinations did when students answered the foundation level of such "building" items incorrectly, but then answered the contingent portion correctly, using the original incorrect answer. Examples 6-4 and 6-5 illustrate short free-response questions.

Essay and Word-Phrase Items. None of the examinations contained either essay or word-phrase items. We considered essay questions (the longest form of extended-response questions) to be single scorable events allocated 20 minutes or more. We considered word-phrase questions (the briefest form of short-response questions) to be those requiring a word or phrase answer, for example, identifying a term for a mathematical concept.

Example 6-4. **(US, AP, Calculus AB, 1993)**
Full Question

25. $\dfrac{d}{dx}\left(2^x\right) =$

(A) 2^{x-1} (B) $(2^{x-1})x$ (C) $(2^x)\ln 2$ (D) $(2^{x-1})\ln 2$ (E) $\dfrac{2x}{\ln 2}$

Example 6-5. **(Sweden, National Examination, 1991)**
Full Question

2. The equation system

$$\begin{cases} 3ax + by = 10 \\ bx - 2ay = 5 \end{cases}$$

has the solution $x = 2$ and $y = -3$. Find a and b.

Practical Activity Items. Given the international push for modeling in secondary school mathematics, one would have expected some items that provide data and require students to generate equations or functions that model them. Such items are the equivalent of the chemistry student having a laboratory practical. However, none of the examinations contained such practical activity items. Had there been an item requiring students to create a model based on data and then employ that model in a decision-making role, then the use of the Practical Activity classification would have been appropriate. However, no such items were observed in the classification of items on the various examinations analyzed. One might argue that the absence of such items is a function of the amount of time available for testing. However, if such connections are a primary goal for mathematics education in secondary school, this goal ought be reflected in the examinations given in the terminal year of secondary schooling.

Use of Diagrams, Graphs and Tables

In general, most of the items on the examinations did not use diagrams, graphs, or tables. (See Figure 6-2.) The French and Japanese examinations did not use these devices at all. Where non-text material *was* used, it was generally in the form of a diagram. For example, the Baden-Württemberg examination used diagrams to illustrate more than 10 percent of its items. The English/Welsh, Swedish, and US examinations used graphs of functions in 3 to 4 percent of their items.

Figure 6-2.   Use of Diagrams, Graphs and Tables in Mathematics Examinations

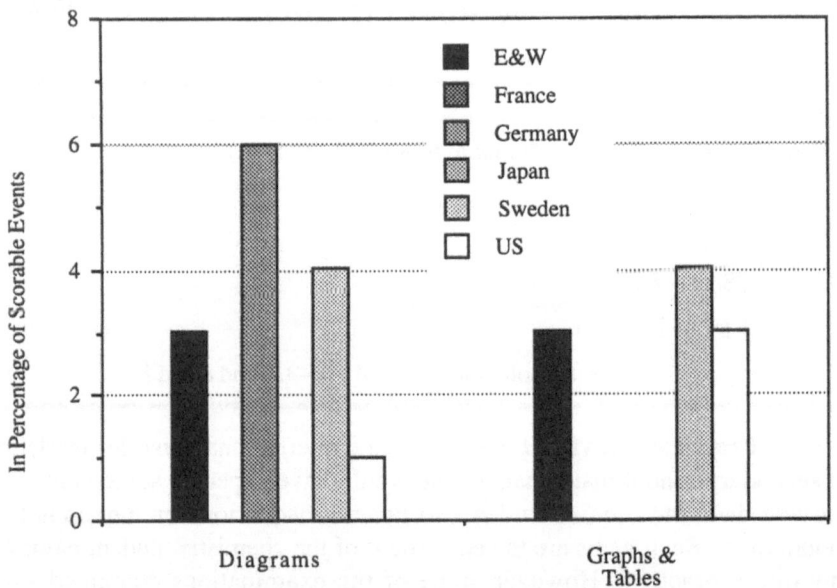

Diagrams. Questions involving diagrams most often were associated with items dealing with geometric content. (See Example 6-6 which includes two scorable events.)

The Japanese examination included an item requiring students to conceive of a complicated relation that necessitated a diagram for comprehension. (See Example 6-7.) It did not provide the diagram for the examinee, rather, the student was left with the task of developing a diagram to show the relationships and solve the puzzle. This item called for students to make a great number of connections, to reason spatially, and to carry the actions in one frame of reference into another. Such items have often been cited as having potential differences in scores based in gender bias between groups of male and female students having equivalent backgrounds in mathematics.

Example 6-6.   **(Germany, Baden-Württemberg, 1992)**
                     Partial Question

The figure given in the Cartesian system of
coordinates represents a solid with vertices
P(3,0,0), Q(0,4,0), R(0,0,0), S(3,0,6),
T(0,4,7), and U(0,0,8). (Measurements are in
cm.) The base area PQR and the top area
STU are planar triangular areas. The side
faces PQTS, QRUT, and RPSU are trape-
zoidal planes.

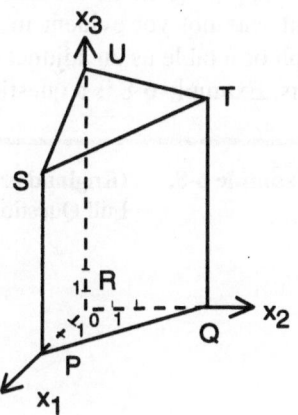

(a) Give a coordinate equations of the plane
     E in which the top area STU lies.
     Calculate the inner angle in the solid
     face between the top area STU and the
     side face PQTS. Determine the angle
     between the edge RU and the top area STU.

Questions b and c follow.

Example 6-7.   **(Tokyo University, 1991)**
                     Full Question

2. Let a, b, and c be positive real numbers. In the xyz-space, consider the
    plane r consisting of points (x, y, z) satisfying the conditions

$$|x| \le a, \qquad |y| \le b, \qquad z = c.$$

    Let P be the source of light on the plane

$$z = c + 1$$

    moving along the ellipse

$$\frac{x^2}{a^2} + \frac{y^2}{b^2} = 1, \qquad z = c + 1$$

    once around. Sketch and calculate the area of the shadow projected by the
    plane R on the xy-plane.

Graphs and Tables.   Graphs and tables are increasingly important adjuncts to textual materials with the advent of technology in the secondary school curriculum and the increased emphasis on working from data to models in applying the mathematics studied in school to real-world settings. This trend was not yet evident in the percentage of questions involving either a graph or a table as an adjunct to text in setting examination questions for students. Example 6-8 is a question involving a graph.

Example 6-8.   (England and Wales, University of London, 1992)
Full Question

7.

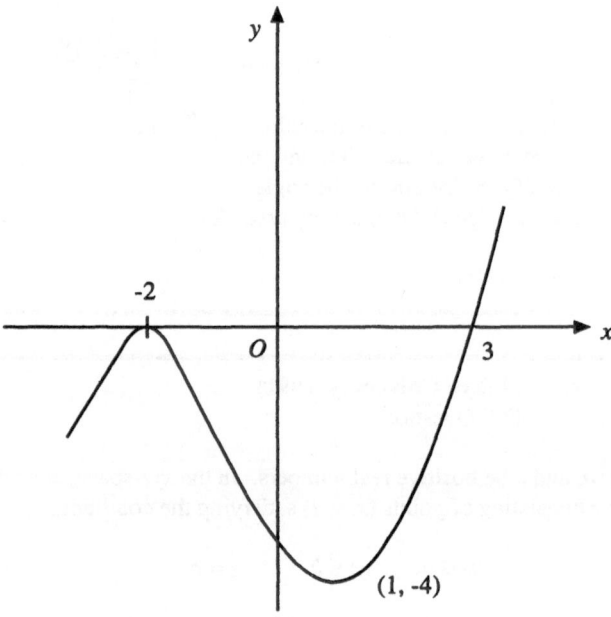

Figure 1 shows a sketch of the curve with the equation $y = f(x)$. This curve has turning points at $(-2,0)$ and $(1,-4)$ and it also cuts the x-axis at the point $(3,0)$. Using separate axes, sketch the curves with the equations
     (a) $y = |f(x)|$,   (b) $y = 4 - f(x)$.

Mark on your sketches the coordinates of the turning points of your curves.

The low level of use of diagrams, graphs or tables compounds the lack of practical items noted previously. This lack of visual or tabular data in the items seems to portray an image of mathematics as a static field of study, dominated by words and symbolic representations. This conservative approach may also represent a misguided view that the use of diagrams may provide some students with more information than others. Such a view is indicative of testing concerns superseding the goals of mathematics education. If students are expected to be able to deal with different representations and work from tabular information, then examinations should present problems involving such information so that we can determine students' mastery of such objectives. Further, if we wish students to develop a rich, dynamic view of mathematics, items involving such information come closer to representing the active use of mathematics in problem solving contexts in the world outside the classroom. The situation portrayed by the data in Figure 6-2 indicates that all of the examinations had a long way to go in moving toward reflecting such objectives.

Examination Topics

This section compares the examinations' topics in two complementary ways. The first subsection notes contrasts only among the most emphasized topics of each examination while the second subsection indicates all mathematics topics covered by each examination. The latter, comprehensive reporting of all topics across all examinations focuses on general topics, such as elementary analysis. The first comparison, dealing with the examinations' top five topics, looks at more detailed topics such as differentiation and integration.[2]

Emphasized, Detailed Topics in Each Examination

Overview of Top-Five Topics. Table 6-3 contains information on the five most emphasized, detailed topics for each of the 10 types of examinations analyzed.[3] No topic appeared as one of the five most emphasized in

---

[2] Appendix A may be helpful for understanding the meaning of the more general topics because it lists the specific topics associated with them. General topics were taken from the Mathematics Framework of the Third International Mathematics and Science Study (Robitaille et al., 1993). The detailed topics were added to permit more specific descriptions of the examinations.

[3] The five most emphasized topics were determined for each examination. These individual examination results were then compiled in Table 6-3 to show all the emphasized topics found in the set of examinations.

Table 6-3.   Mathematics Topics, Detailed: Five Most Emphasized in Each Examination

Percentage of Examination Score

	AEB	UOL	Aix	Paris	Bav	B-W	Jap	Swe	US-B	US-C	Topic Frequency
Applications of Definite Integral	6				11			26	15	13	5
Applications of Derivative	5	5				4	8	12	15	7	7
Interpreting Function Graphs			11	16	9	7					4
Derivatives: Max & Min		11							11	7	3
Trigonometric Equations & Identities	5	8					8				3
Conditional Probability					6	8					2
Probability Distributions		3			11	5					3
Polar Coordinates	15	4									2
Numerical Probability							8				1
Limits at Infinity						4					1
Fundamental Theorem of Calculus			5						5		2
Higher Order Derivatives				6							1
Logarithmic and Exponential Equations								10			1
Geometric Transformations				10	4						2
Properties of Functions										6	1

Table 6-3.   Continued

	AEB	UOL	Aix	Paris	Bav	B-W	Jap	Swe	US-B	US-C	*Topic Frequency
Coordinates in 3-space							8	6			2
Equations of a Line in Plane							8				1
Right Angle Trigonometry								**10**			1
Polar Coordinates	6	5									2
Measurement of Area							8				1
Circles and Their Properties					7						1
Geometric Dilations				8							1
Planes and Lines in Space							8				1
Ellipses					7						1
Mathematical Induction							8				1
Derivatives of Implicit Functions									5		1
Antiderivatives									7		1
DeMoivre's Thm. & Complex No.			5								1
** Topic Concentration	37	26	40	46	41	28	56	66	52	39	

Bold indicates values of 10 percent or greater.

Blank cells only indicate that the topic was not one of the top five in the examination. The examination still may have included the topic.

* Topic Frequency indicates the number of examinations having the topic.

** Topic Concentration indicates the percentage of the examination comprised by its five most emphasized topics.

every examination! Three of the most frequent topics were aspects of elementary analysis: applications of the definite integral, applications of the derivative, and the relationship between derivative values and maxima and minima of functions. The other two most emphasized topics were about funcions—interpretation of function graphs, and trigonometric equations and identities. These most frequent topics, which occurred in 3-7 examinations, are consistently found in solid courses in precalculus, elementary functions, or the first year of calculus. One other topic was emphasized on these examinations—probability distribution. Since the other 22 topics in Table 6-3 were emphasized only in one or two examinations, the examinations' differences in emphases were greater than their similarities.

Topic Concentration. The summary row of Table 6-3 indicates the examinations' topic concentration. It shows the percentage of each country's examinations comprised by their five most emphasized topics.[4] The Japanese, Swedish and US-AB examinations addressed fewer topics than most, the majority of examinations had moderate topic concentration of 39-46%, and the UOL and Bavarian examinations had broader topic coverage. The UOL examinations covered the most topics, having a topic concentration of only 26 percent. Because the Japanese and Swedish examinations had so few questions (see table 6-1), their most emphasized topics comprised 56 to 66 percent of the examinations. Since the Calculus AB course of the Advanced Placement course is limited to calculus, it is understandable that the US-AB examination also had a high topic concentration (52%).

All General Topics Covered by Each Examination

Table 6-4 shows the proportions of each examination devoted to eight broad mathematics categories that span the field of mathematics addressed by this study's set of examinations. This analysis is designed to fit the categories of the TIMSS Mathematics Curriculum Framework, while still detailing the relative weighting given the items in the examinations considered as a whole. The following are more detailed descriptions of the topics in Table 6-4: geometry-position, visualization, and shape forms; functions, relations, and equations; elementary analysis; data representation, probability, and statistics; geometry relations-symmetry, congruence, and similarity; proportionality; numbers; and measurement. The first three of these categories are "universal topics"; the remaining five are "less common."

---

[4] This statistic can be thought of as an indicator of an examination's breadth in topic coverage. Examinations with higher and lower topic concentrations focused on fewer or more topics, respectively.

Table 6-4.   Mathematics Topics, General: All Topics Found in Each Examination
Percentage of Examination Score

Country/ Exam	Cluster	Universal Topics				Less Common Topics			
		Geo-metry - Forms	Func-tions	Analy-sis	Data	Geo-metry - Relation	Pro-portion-ality	Numbers	Mea-sure-ment
E&W - AEB	1	34	39	23	—	—	1	2	1
E&W - UOL	1	15	33	27	17	1	—	6	1
Fr. - Aix	1	15	18	41	—	17	—	6	3
Fr. - Paris	1	27	30	26	—	13	—	3	1
Ger. - Bav.	2	22	15	21	36	—	—	—	6
Ger. - B-W	2	24	11	18	35	9	3	—	—
Japan	2	29	14	28	17	—	—	4	8
Sweden	3	6	28	52	—	—	10	—	4
US - AB	4	—	10	90	—	—	—	—	—
US - BC	4	1	6	93	—	—	—	1	—

In analyzing the data to discover similarities among the various countries' examination topics, more differences than commonalities were found. Geometry-forms, functions, and analysis were the only main topics included in all or most examinations. The greatest variance in emphasis among these universal topics was that the Swedish and US examinations strongly downplayed geometry-forms while emphasizing analysis.

Some interesting topic differences appeared within the countries in this study having multiple examination regions or boards (England and Wales, France, and Germany). A majority of the topic coverage within a country was similar; in the German examination, for example, each of the four most emphasized topics only varied by 2-4 percent between states. However, within country differences were still notable.

The AEB examination emphasized geometry-forms and functions, and the UOL examination stressed functions and data analysis. While both the Aix and Paris examinations stressed geometry-relations, the former gave more emphasis to analysis and the latter to functions and geometry-forms.

Among the minor topics in the German examinations, B-W included geometry-relations and proportionality, whereas Bavaria included measurement.

Example 6-9 focused on students' ability to determine an expression for the measurement of a pyramid having one of its vertices expressed in parametric terms. It required students to make a number of connections between the standard measurement formulas and their knowledge of distance between points and planes in space. Alternatively, the student could approach the item through matrices.

Example 6-9.   (Germany, Bavaria, 1992)
Partial Question

In a Cartesian coordinate system, points A(8, 0, 0), B(8, 3, 0), $C_t(4t + 5,\ 3,\ -3t)$ where $t \in R$, and D(0, 0, 6) are given.

Question 1, a-c precedes.

2. (a) Prove that the volume of pyramid ABC_tD is 9. Give a geometric proof that the volume is independent of t.

   (b) Determine the distance d of point C_t from the plane ABD.

Question 2c follows.

The data on topics across countries were examined using cluster analysis, a statistical approach designed to identify groups (clusters) of countries having patterns of common topic emphasis. This process identified four clusters of countries as having common topic emphasis/usage patterns, even though differences still exist between specific topic choice in individual clusters. The clusters and their global patterns of topic emphasis were:

- Cluster 1, the AEB, UOL, Aix, and Paris examinations, elevated the importance of geometry-forms, geometry-relations, functions, and analysis as topics to examine. While the examinations in this cluster held these topics among those deemed more important, they differed in that the English/Welsh examinations tended to focus more on the application of this material to the physical sciences while the French examinations tended to examine the material in a more abstract fashion, reflecting a "pure" mathematics approach. It is interesting to note that the countries in this cluster attach high stakes to their examinations for college entrance.

- Cluster 2, the two German and the Japanese examinations, stressed geometry-forms and data analysis, while downplaying analysis. These examinations were a bit more similar in nature, but they differed in length, number of items, and in level of difficulty. The German examinations seemed to reflect more application of work studied in the classroom, while the Japanese examination tended to approach the topics through items that were more indicative of real problem solving in"pure" settings. The countries in this cluster also place high stakes on their examinations.

- Cluster 3 consisted solely of Sweden's examination. It stressed proportionality, functions, and analysis, while downplaying geometry-forms. The items on the Swedish test were perhaps the most reasonable for a wide range of students. The nature of the Swedish examinations is that of a low-stakes examination for individual students.

- Cluster 4 consisted of the two US Advanced Placement calculus examinations. These placed an extremely high emphasis on analysis to the exclusion of all other topic areas except functions. These examinations also differed from the examinations in other clusters in that they are used for proficiency credit and not for college admission in most cases.

Overall, two content areas—functions, relations, and equations and elementary analysis—represent the strongest areas of commonality across the various examinations studied. Further analysis of these content areas follows. However, it is also interesting to note that the content clustering paralleled the usage of tests across the various countries to partition students based on their levels of performance. Is it the case that such topic choice may be more related to examination purposes (high-stakes versus low-stakes) of discriminating between pupils than developing a comprehensive profile of what the student knows and is able to do?

Functions, Relations, and Equations

Table 6-5 is a breakdown of each examination's proportion of total points that were allocated to two subtopics within functions, relations, and equations.

The 10 examinations fell into two distinct clusters. The first cluster—consisting of the two English/Welsh examinations and the examinations from Paris and Sweden—devoted between 28 and 39 percent of their total score to items from the overall function category. The Aix, German,

Table 6-5.   Coverage of Functions Topics

Country/ Examination	*Total Coverage of Functions (Percentage of Examination Score)	Patterns, Relations, & Functions (Percentage of Function Topics)	Equations & Functions (Percentage of Function Topics)
Cluster 1*			
E&W - AEB	39	20	**80**
E&W - UOL	33	17	**83**
Fr. - Paris	30	**71**	29
Sweden	28	14	**86**
Cluster 2*			
Fr. - Aix	18	**68**	32
Ger. - Bav.	15	**82**	18
Japan	14	**80**	20
Ger. - B-W	11	**92**	8
US - AB	10	**84**	16
US - BC	6	28	**72**

Examinations ordered by total coverage of functions topics. Bold indicates most emphasized functions topic in each examination.
* Clusters 1 and 2 had 28-39 and 6-18 percent of examination score allocated to functions topics, respectively.

Japanese, and US examinations constituted cluster 2; they devoted from 6 to 18 percent of their overall value to the topic of functions.

Cluster 1 was more procedurally oriented, focusing on equations and functions, while cluster 2 examinations focused much more on the theoretical and conceptual aspects of the topic—patterns, relations, and functions. The Paris and US BC Calculus examinations were outliers to their two clusters relative to this comparison. Sample questions 6-10 and 6-11 illustrate the topic difference.

Example 6-10.   **(England and Wales, University of London, 1991)**
Full Question

2. Given that $|x| \neq 1$, find the complete set of values of x for which

$$\frac{x}{x-1} > \frac{1}{x+1}.$$

Example 6-11.   **(US, AP, Calculus AB, 1988)**

1. Let f be the function given by $f(x) = \sqrt{x^4 - 16x^2}$.

   (a) Find the domain of f.

   (b) Describe the symmetry, if any of the graph of f.

Elementary Analysis

Table 6-6 shows the proportion of total points on each examination that was allocated to four more specific topics within elementary analysis: limits continuity, sequences, and series; differentiation; integration; and other topics. "Other" topics were composed mainly of items dealing with multiple integration or elementary differential equations.

   The examinations could be grouped into two clusters based on their coverage of elementary analysis content. The first cluster consisted of the two US examinations, both of which devoted over 90 percent of their point values to the category of elementary functions; the second cluster consisted of all the other examinations. This distinction notwithstanding, there is little difference in the overall distribution of points in either cluster. On average, the examinations devoted 10 percent of their analysis points to limits, continuity, sequences, and series; 50 percent to differentiation; 35 percent to integration; and 5 percent to other topics. There were some notable exceptions, however. The Baden-Württemberg and Swedish examinations had no items on limits, continuity, sequences, and series. The Paris examination had no items on integration. And the only countries with content from the "other" area were England/Wales, Japan, and the United States. More detailed findings on some of the elementary analysis topics follow.

Table 6-6.   **Coverage of Elementary Analysis Topics**

Country/ Exam	Total Coverage of Elementary Analysis (Percentage of Examination Score)	Limits Continuity, Sequences, & Series	Differen- tiation	Inte- gration	Other
		(Percentage of Analysis Topics)			
US - BC	**93**	13	48	35	4
US - AB	**90**	5	60	34	1
Sweden	52	—	35	65	—
France - Aix	41	20	52	28	—
Japan	28	15	45	10	30
E&W - UOL	27	27	36	30	7
Fr. - Paris	26	22	78	—	—
E&W - AEB	23	15	37	43	5
Ger. - Bav.	21	20	54	26	—
Ger. - B-W	18	—	28	72	—
Average	42	10	50	35	5

Examinations ordered by total coverage of analysis topics. Bold indicates examinations with greatest emphasis on elementary analysis topics.

Differentiation. Each examination somewhere asked about the overall nature of the shape of a curve or for a description of its overall patterns of variation. (See Examples 6-12 and 6-13.)

The coding classifications representing at least 9 percent of the scorable events in the differentiation topic were:
- derivatives of sums, products, and quotients (9 percent);
- higher-order derivatives (10 percent);
- analyses of the relationship between the derivative and maxima/minima (17 percent); and
- applications of the derivative (17 percent).

Example 6-12.   **(England and Wales, Associated Examining Board, 1991)**
                    Partial Question

9. Figure 1 shows a sketch of the curve defines for $x > 0$ by the equation

$$y = x^2 \ln x.$$

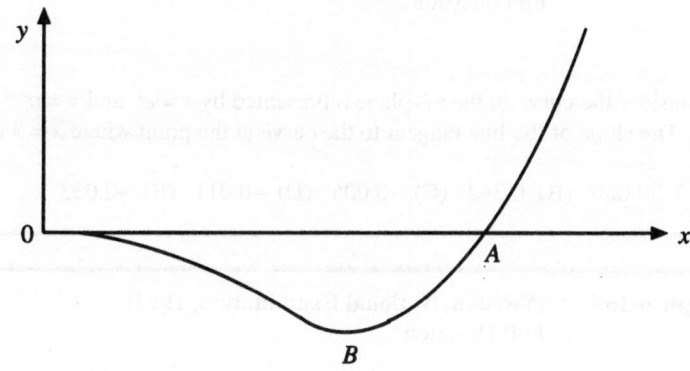

The curve crosses the x–axis at A and has a local minimum at B.

(a) State the coordinates of A and calculate the gradient of the curve
    at A.

(b) Determine the coordinates of B and determine the value of $\dfrac{d^2y}{dx^2}$
    at B.

Question 9c follows.

Example 6-13.   **(France, Paris, 1992)**
                    Partial Question

1. Given that h_n is the numerical function defined on $]-1, +\infty[$ by:

$$h_n(x) = n \ln(1 + x) + \frac{x}{1 + x}.$$

Study the direction of variation of h_n.

Question 1 continues.

The distinction between the last two topics was often whether the questions had context or not. (See Examples 6-14 and 6-15.) Unlike the previous grouping of analysis questions, the expectations of students in derivation of power functions were generally shared by the examinations containing differentiation events.

Example 6-14.   **(US, AP, Calculus BC, 1993)**
Full question

25. Consider the curve in the xy–plane represented by $x = e^t$ and $y = te^{-t}$ for $t \geq$ 0. The slope of the line tangent to the curve at the point where $x = 3$ is

    (A) 20.086 (B) 0.342 (C) –0.005 (D) –0.011 (E) –0.033

Example 6-15.   **(Sweden, National Examination, 1992)**
Full Question

8. A patient is given an intravenous injection of a certain substance. At the beginning the concentration of the substance is 3.00 mg/ml. After 12 minutes the concentration has decreased to 2.38 mg/ml. The substance is active as long as the concentration exceeds 0.80 mg/ml. How long will such an injection be active, if the concentration decreases exponentially?

While there was a great detail of variability across examinations with respect to the ways the concept of the derivative and its applications were assessed, all of the examinations asked students to exhibit their ability to apply the power rule for differentiation in rather straightforward settings.

Integration. The scorable events dealing with integration were also widely spread across the various subtopics. Two subtopics—basic integration formulas and applications of the definite integral—accounted for 9 percent and 46 percent, respectively, of the coding. The applications universally noted were areas under or between curves and volumes of surfaces of revolution. (See Example 6-16 and the first question in Example 6-17.)

Other Elementary Analysis Topics. The "other" topics under analysis area consisted of items that dealt with extensions of the limits-differentiation-integration curriculum. (See Example 6-18 and the second

Example 6-16.   **(Germany, Baden-Württemberg, 1992)**
Partial Question

For every $t > 0$ the function ft is given by

$$ft\,(x) = \ln(x^{\,2} + t\,) \ \text{ for } x\,\varepsilon\,\text{R}.$$

Let Kt be its graph.

Questions a and b precede.

(c) $K4$ and the line $y = \ln 8$ enclose a surface. Calculate the volume of the solid formed by rotating it about the y – axis.

Example 6-17.   **(England and Wales, Associated Examining Board, 1992)**
Partial Question

10. The tension in a light elastic spring is given by Hooke's law when the extension of the spring is x meters, where $0 \le x \le 0.5$. When $x = 0.5$ the tension is 7.35N. Find the work done in increasing x from 0.2 to 0.5.

   For $x > 0.5$ the tension is not given by Hooke's law. Values of the tension T Newtons, for specific values of x in the range $0.5 \le x \le 0.9$ are given in the following table.

x	0.5	0.6	0.7	0.8	0.9
T	7.35	9.03	10.80	12.90	15.03

   Using Simpson's rule, with five ordinates, show that the work done in increasing x from 0.5 to 0.9 is approximately $4.39J$.

---

question in Example 6-17.) Such questions only accounted for 3 percent of all analysis scorable events. Four of these seven scorable events dealt with first-order linear differential equations, two with numerical considerations to integration, and one with multiple integration.

The fact that there were only seven events in this category suggests virtually no extension beyond the limits-differentiation-integration curriculum in the countries whose examinations were analyzed.

Example 6-18.   (US, AP, Calculus, AB 1988)
Full Question

38. For $x > 0$, $\int \left(\dfrac{1}{x} \int_1^x \dfrac{du}{u} \right) dx =$

(A) $\dfrac{1}{x^3} + C$ (B) $\dfrac{8}{x^4} - \dfrac{2}{x^2} + C$ (C) $\ln(\ln x) + C$

(D) $\dfrac{\ln(x^2)}{2} + C$ (E) $\dfrac{(\ln x)^2}{2} + C$

Performance Expectations

Figure 6-3 illustrates the categories of performance expectations emphasized by the various mathematics examinations. While the TIMSS Curriculum Framework contained five main categories of performance expectations for Mathematics, as seen in Table 1-3, only three types of performance expectations occurred in most of the examinations. These were Mathematical Reasoning, Investigating and Problem Solving, and Using Procedures. Across countries, they accounted for 97 percent or more of the performance expectations expressed in any examination.

The results of a cluster analysis to determine patterns across examinations gave the three groupings illustrated in Figure 6-3. The examinations in cluster 1 were noted by the relative balance between Using Routine Procedures and Investigation and Problem Solving, while still having some Reasoning. The examinations in cluster 2 loaded on Investigation and Problem Solving more heavily than Using Routine Procedures, with little attention given to Reasoning. Cluster 3 relied predominantly on Using Routine Procedures (70 percent), with the remainder split almost 2 to 1 between Investigation and Problem Solving and Reasoning.

Thus examinations from AEB, Japan, Sweden, and US-BC put relatively equal emphasis on using routine procedures and problem solving. Examinations from UOL, Aix, Paris, B-W and Bav. stressed problem solving over the knowledge of routine procedures, and the US-AB examinations put heavy emphasis on performing routine procedures or algorithmic processes.

Figure 6-3.   General Performance Expectations in Mathematics Examinations

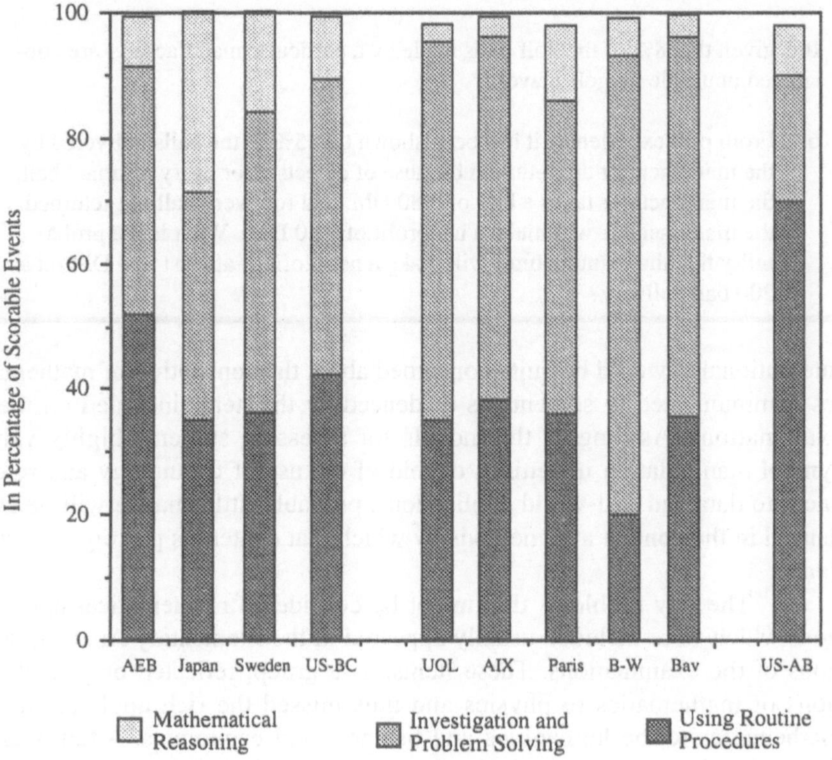

Mathematical Reasoning · Investigation and Problem Solving · Using Routine Procedures

(Y-axis: In Percentage of Scorable Events; categories: AEB, Japan, Sweden, US-BC, UOL, AIX, Paris, B-W, Bav, US-AB)

Summary

Overall, the examinations—except for some mathematical physics items on the English/Welsh examinations seen previously and a few stochastics items from the German examinations (see Example 6-19)—were devoid of connections between mathematics and real-world contexts.

This lack of connection to relevant applications of mathematics expresses the prevailing conception of school mathematics as a subject characterized by recognition/repetition of definitions and theorems and the performance of symbol manipulation procedures. Little in the mathematics sections of the examinations connected them to the richer, problem-solving vision for school mathematics described in the National Council of Teachers of Mathematics' *Curriculum and Evaluation Standards for School Mathematics* (NCTM, 1989). On this basis alone, mathematics educators,

Example 6-19.   **(Germany, Bavaria, 1992)**
Partial Question

It is given that 8% of the golf balls made by a particular manufacturer are considered unusable by golf players.

6. From past experience, it has been shown that 5% of the balls delivered by the manufacturer are returned because of defects. For every returned ball, the manufacturer takes a loss of 0.80 DM, and for every ball not returned the manufacturer will make a net profit of 1.20 DM. What is the probability that the manufacturer will make a net profit of at least 210 DM on a 200 ball delivery?

internationally, would be quite concerned about the conception of mathematics communicated to students as evidenced by the items included on these examinations. As long as the models for assessing students highly value symbol manipulation in settings devoid of the use of technology and references to data and real-world applications, probably little change will be evidenced in the content and methods by which that content is portrayed to students.

The few problems that might be considered mathematical applications which were included usually appeared in the Elementary Analysis sections of the examinations. These items, as a group, reflected only applications of mathematics to physics and thus missed the rich applications of mathematics to the humanities and business and economics. Additionally, the applications to the physical sciences often reflected the types of items marked for deletion from the curriculum by those involved in current curricular reform efforts. One such example is the ubiquitous volumes by solids of revolution items found in almost all of the examinations. This item type is the very kind of template exercise being targeted for elimination in current reform efforts in the teaching of calculus.

Even a cursory study of the examinations would reflect the differing purposes and curricula that the nations have for school mathematics. However, it was surprising to note that the cluster analysis results for examination topics paralleled the varying purposes of the examinations for sorting and selecting students. The Japanese examination was designed to rank and identify superior students, while not necessarily measuring their command of the entire curriculum. The US Advanced Placement examinations in calculus, on the other hand, were designed to certify that students have mastered the minimal concepts and skills required to award university-level placement and credit. The English/Welsh examinations covered a broad range of topics and reflect expectations of student mastery of concepts from physics as well as mathematics. The French examinations were much more focused in the belief that performance on a few well-chosen items will reflect student competence on a broader scale. This is somewhat the same approach taken by the Swedish examinations, but the level of difficulty of the items was much lower, as the examination was not used to sort students for advancement, but rather build a basis for norming mathematics performance at the national level. The German examinations, at least for the two states analyzed, were perhaps the most closely tied to the local curriculum, being developed at the state/local level to ascertain student mastery of mathematics.

Secondary students in all countries were sufficiently challenged by their respective countries' examinations in mathematics. However, in terms of overall difficulty, the examinations fell into three general groupings:

- The first, and easiest, was the US AB Calculus examination focusing on elementary functions and basic integration and differentiation. This characterization matches the intent of the examination. The US AB Calculus examination, as a placement/proficiency examination for a semester of calculus, had less breadth than many of the other examinations.

- The next grouping consisted of both French examinations, both German examinations, both English/Welsh examinations, the Swedish examination, and the US BC calculus examination. These examinations dealt with a broad range of topics, and were set by some national body focusing on making a distinction between either students or levels of marks.

- The Japanese examination appeared to be the most difficult. It is a university entrance examination aimed at separating out the most capable students.

Part II

Understanding the
Examination Systems

Commentary

Jack Schwille

ONE OF THE VERY important things this project has done is to show clearly that content analysis of external examinations in another country does not take one far enough in development of standards. Gathering copies of examinations and giving them to authorities in subject matter does not alone yield a definitive assessment of standards in other countries that might serve to guide curriculum policy making in the United States or some other country.

Examinations have to be put in context. What seems to be the case is that when subject-matter experts speak of the rigor of an examination they have reviewed, they do it with a hypothetical reference norm group of students in mind—the sort of students they themselves have experienced—and with assumptions about the curriculum, the opportunities for learning to which these hypothetical students have been exposed. In other words, the same examination item may be judged impossibly hard for students with no exposure to the topic being examined, very difficult for those with minimal exposure, but perhaps fairly easy for those who have had a great deal of teaching and examination coaching for items of this type.

My conclusion from this is not that we should abandon the analysis of examinations from other countries, but that we should become more open and explicit about the assumptions we are making. We should, perhaps, approach the review of such examinations as if we were doing a book review, analyzing the examinations' strengths and weaknesses as we perceive them, while making our own assumptions, point of view, and background explicit.

There are advantages of a cross-national approach, however. We do have taxonomies of content within which these examinations of different purpose and consequences have been placed. In a sense, these taxonomies themselves constitute an international context which should be taken into account, and our understanding of what is being achieved in a particular national context will henceforth be colored by our understanding of what is being tested in one country and not in another. For example, the analysis of the chemistry examinations includes a number of what seem to be very important content differences among countries. One is the lack of emphasis in the United States on organic chemistry and biochemistry.

Currently, there are two opposing approaches to the development of international understanding of curriculum. One is represented by the analysis of curricula across countries to identify commonalities and differences in content and performance expectations, using a common taxonomy or framework for the analysis. The analysis of where a nation is located within this structure of commonalities and differences contributes to the development of standards. The most relevant existing work to this approach is in the International Association for the Evaluation of Educational Achievement (IEA) curriculum analysis as currently being applied to the Third International Mathematics and Science Study (TIMMS). However, the TIMSS curriculum analysis—like the research effort discussed in this volume—did not have the development of standards as its aim.

The other approach shuns the cross-national analysis of commonalities and differences, and tries to understand what standards mean on a nation's own terms. This approach is being tried by the New Standards project and is discussed in Resnick, Nolan, and Resnick (1994). It involves addressing benchmarking questions such as "What are students in other countries expected to know and be able to do at key transition points in their schooling careers?" The analysis described in this book can be viewed from both of these perspectives.

The analysis of these examinations should increase the demand for more comparative research that will put the examinations into the context of students' overall opportunities to learn and their consequences for student learning and subsequent achievements. We need more empirical longitudinal insight into what sense students make of these types of subject-matter before, during, and after their preparation for critical examinations. Such a research agenda would push us into much needed comparative empirical research on the postsecondary learning careers of students, a very neglected area.

It is really amazing to think how much comparative research on external examinations remains in its infancy. Work such as we have been discussing here represents research that is long overdue. One wonders why it has not been done before. And this is but one of many areas where one cannot adequately understand what is being done in American education without understanding what is being done in other countries. Before we set supposedly unique national criteria and goals for our own educational system, we would benefit from careful investigation of how typical or unusual we are in existing educational practice.

7

Comparing Examination Systems

Edward Britton
Simon Hawkins
Matthew Gandal

TO MAKE VALID INFERENCES about the examination comparisons in part I, and avoid unwarranted ones, the context of the examinations should be understood. Chapter 1 provides brief overviews of each country's examination system and Chapter 8 describes them in detail. The following chapter attempts to illuminate the most notable contrasts and similarities among the seven examination systems in this study. It compares examination purposes, numbers of examinees and pass rates, governance, programs of study leading to examinations, examination creation and scoring, and system and student costs. Chapter 7 begins with highlights of the cross-country findings and closes by noting both known differences among the examination systems and some information needs that make comparing the difficulty of examinations problematic.

Highlights

- The great majority of college-bound students in countries other than the United States *must* take and pass some advanced subject-specific examinations.

- The educational systems of England and Wales, France, and Germany are now preparing increasing numbers of students for these examinations, rather than targeting an elite few as they did in the past.

- In France, Germany, and Israel, academically oriented students who do not seek further education still take these examinations because passing them is a prestigious credential in their societies.

- While only 6.6 percent of US students take Advanced Placement (AP) examinations, roughly a quarter to a half of all students in other nations take and pass advanced subject-specific examinations.

- National examinations, often assumed to be a single nationwide examination, sometimes were sets of examinations from different regions or examination boards.

- In England and Wales, France, Germany, Israel, and Sweden, most college-bound students pass these challenging examinations because they are in formal academic tracks designed in part to prepare them for the examinations.

- In England and Wales and France, examinations administered at the end of lower secondary school determine whether students will follow an academic or vocational program; in Japan, entrance examinations are used by high schools to admit students of similar achievement to a given school.

- The subject concentration of courses of study leading to examinations varies across countries. Students in England and Wales focus heavily on only three subjects, while French students typically take examinations in seven or eight subjects.

- Some educational systems provide alternatives to passing examinations only at the end of secondary school. These alternatives are second chances to take examinations after additional preparation, routes to college that do not involve these examinations, and opportunities to take examinations a number of years after school completion.

- In most countries, committees of select teachers and university faculty create and score examinations. In many German states, however, the teachers of the students taking the examinations are responsible both for creating and scoring them. In Japan, university faculty have almost complete responsibility for these tasks.

- Knowing the costs of other nations' examination systems is critical for considering the implications of these systems in the US context, but little specific cost information is available.

- Examinations are free for students in England and Wales, France, Germany, Israel, and Sweden. The costs of taking them in Japan and the United States may deter access by students from less advantaged families.

- Cross-country comparisons of the difficulty of examinations is problematic because of known differences in examination systems and inadequate information about other pertinent factors. Differences exist in students' total examination load and the relative weights given to examination grades and course grades. More information is needed about rigor of scoring examination items and assigning cutoffs for total grades, curricular links between instruction and examinations, in-school coaching for examinations and out-of-school examination preparation.

Examination Purposes

In each country except the United States, college-bound students seeking to study in a university must pass demanding subject-specific examinations (see Table 7-1). In France, Germany, and Israel, even many students who do *not* go on to college take these examinations because they are a prestigious credential in their societies. In France, for instance, the baccalauréat course of study should not be seen strictly as a college track because it is generally

Table 7-1.   Functions of Advanced Subject-Specific Examinations

Country	Prestigious - Secondary Credential	Required - University Admission	Optional - University Credit
E&W		X	
France	X	X	
Germany	X	X	
Israel	X	X	
Japan		X	
Sweden		X	
US			X

accepted that without passing the baccalauréat, nonvocational jobs are hard to find. Thus, in 1990, 34 percent of the age group passed the general (academic) baccalauréat, but only 11 percent of the age group went on to university or other advanced study (Langlois, 1991).

While New York and California administer voluntary examinations that may influence university admission, there are no advanced subject-specific US examinations that all students nationwide must pass in order to be eligible for university study. The advanced subject-specific examinations that *do* exist are voluntary. Students can sometimes obtain college credit for their scores on Advanced Placement examinations.

Number of Examinees and Pass Rates

One of the most striking findings of our research is how many students take and pass advanced subject-specific examinations at the end of secondary school. Approximately one-third to one-half of the age cohort in England and Wales, France, Germany, Israel, Japan, and Sweden take advanced subject-specific examinations such as the ones described in this volume (although not necessarily in science or mathematics). (See Figure 7-1.) In sharp contrast, only 6.6 percent of US 18-year-olds take at least one AP examination.

Note, however, that not all of an age cohort within an individual country take and pass their examinations in the same year. In most countries, this variation is due to two factors:

- retaking the examination after an initial failure, a practice prevalent in England and Wales, France, and Japan

- retention, or holding back a student in a particular grade

French and German students in any track, whether vocational or academic, may be held back for a year but remain in their track. Retention is a tool used to ensure better preparation for examinations. In the United States, generally only nonacademic students are held back. Even so, most of an age cohort in the United States and Japan tend to leave secondary school at the same age, unlike students in France and Germany (see Table 7-2).

Until recently, the high standards reached by students in other nations could be somewhat offset by the claim that their systems were elitist and prepared relatively few students for college. This perception, however, is increasingly difficult to justify. Several nations have made and are making considerable efforts to bring a greater proportion of their age cohorts up to the level of performance required for these examinations. Today, close to

Figure 7-1.   Percentage of Age Cohort Who Take and Pass Advanced Subject-Specific Examinations[1]

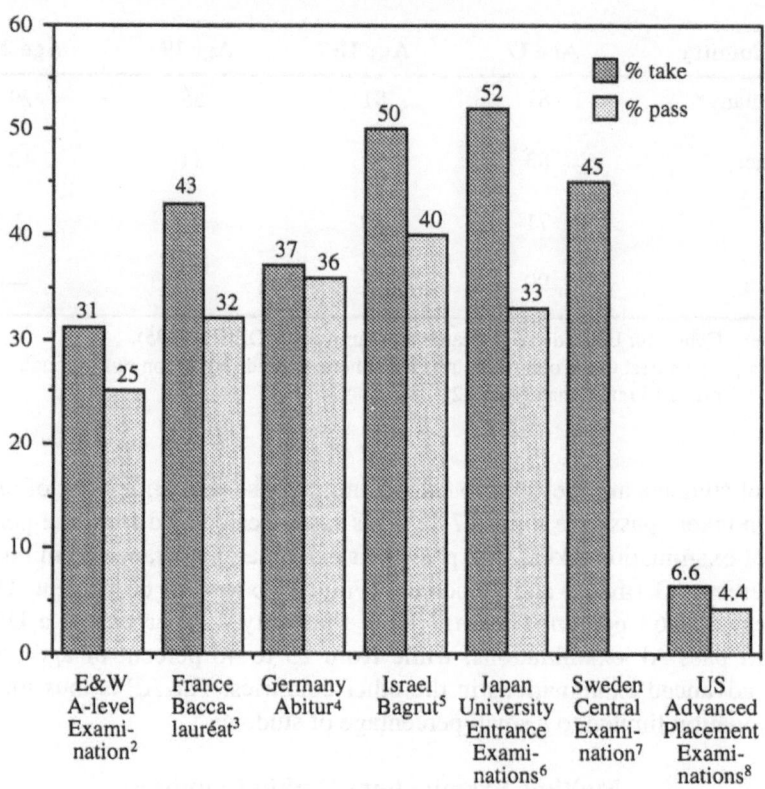

1. Age cohort refers to the percentage of an age cohort who eventually take and/or pass the examinations in an academic course of study, although they may not all take or pass it in the same year, for instance, students may fail the baccalauréat once, study for a year, and then take it again and pass it.
2. A-level candidates generally take three examinations. Although 25% of the age cohort passes at least one examination, only 15% of the cohort passes three examinations. (Evans, 1994)
3. These data represent only those students who received a general (academic) baccalauréat, which includes a set of at least six examinations in different subjects. Students in non-academic courses of study also take baccalauréat examinations. (Bodin, 1994)
4. These data represent students who received the Abitur, which includes a set of four examinations in different subjects. (Stevenson et al, 1994)
5. These data represent students who received the Bagrut, which includes a set of from five to ten examinations in different subjects. (Tamir, 1995)
6. These data represent the number of students who apply to university (and thus take the university entrance examinations) and are eventually accepted. (Ministry of Education, Science and Culture, 1994; Centre for Educational Research and Innovation, 1993)
7. Students do not pass or fail the Central Examination, but have its score factored into their final course grades. The data here represent only those students who follow an academic/theoretical line. Students in non-academic courses of study also take examinations. (Gisselberg, 1992)
8. Most students take one or two examinations. AP examinations are available only in about 50% of US high schools. (Advaned Placement Program, 1993; National Center for Education Statistics, 1993)

Table 7-2.   Percentage of Age Cohort in Secondary School from Ages 17-20

Country	Age 17	Age 18	Age 19	Age 20
Germany*	81	81	55	29
France	83	58	34	12
US	71	20	6	2
Japan	90	2	—	—

Source: Centre for Educational Research and Innovation, OECD (1995).
*Note: In contrast to the other countries listed here, German education runs through
          grade 13 rather than grade 12.

half of students are prepared to take examinations—and a majority of examination takers pass (see Figure 7-1). Pass rates—calculated from the percentage of examination takers who pass their examinations—range from highs of 95 percent (Germany) and 75 percent (France) to lows of 66 percent (United States) and 63 percent (Japan).[1] Note that only 4.3 percent of a US age cohort pass AP examinations, while from 25 to 40 percent of age cohorts pass advanced examinations in the other countries. The AP is thus the only examination limited to a small percentage of students.[2]

Multiple Examinations Within Countries

Having a national examination system does not necessarily mean that a particular country has a single, nationally developed and administered examination. The federal governments in Sweden and Israel and the College Board in the United States do develop examinations for their respective countries as a whole. But in France and Germany, students in various parts of each country take examinations that different governmental organizations develop according to a national specification. In England and Wales, schools choose examinations from a number of competing private examination boards. For these

[1] All rates are national averages and generally are typical for any group of college-bound students. The exception is Japanese pass rates; these vary considerably by university, thus making the Japanese pass rate less typical.
[2] Note too that about half of all high schools in the United States do not offer AP examinations.

countries, the "national" examination is actually a set of comparable examinations.

In England and Wales and in France, the national ministries of education have considerable control over the content and difficulty level of the examinations, though their development and administration are somewhat decentralized. In Germany, the national ministry plays a less prominent role in the development of the Abitur. The ultimate responsibility for the examinations lies with each German state (Land). Supervision is provided by a committee of education ministers from each of the Länder.

While Japan does have a set of national examinations—the University Entrance Center Examination, produced by university faculty on behalf of the Ministry of Education, Science, and Culture—only a few of the universities require students to take them. Instead, each university, or department within it, exercises autonomy by developing its own set of entrance examinations. The content of these examinations, however, does reflect the national curriculum. Only the universities' individual entrance examinations are required of all students aspiring to higher education.

Formal Differentiation Between Academic and Nonacademic Students

The examinations in this book are taken by secondary school students who have decided to continue their education in colleges and universities. Students in each country make that choice at different ages, and the degree to which they are separated from and involved in different course work than non-college-bound students varies from country to country (see Table 7-3).

The national curricula in England and Wales, France, and Japan ensure that all students up to a certain age are exposed to a common core of subject matter. In France, differentiation occurs at age 15; in England and Wales, partly at age 14, and partly at age 16. In these countries, students who want to continue in school choose between college-preparatory and vocational tracks, and most take a set of examinations prior to entering these tracks.

In Japan, differentiation of another sort occurs at the end of lower secondary school. At this point, students enter a competitive admissions process (complete with examinations) to attend the high schools of their choice. Though the national curriculum is taught at all high schools, a clear hierarchy of quality exists, and some schools have higher concentrations of college-bound students than others.

Formal tracking takes place at the earliest age in Germany. In most Länder, when children have completed four years of school—usually at age 10—teachers and parents decide among three secondary tracks, one of which prepares students for university study. Each of the three tracks has its own separate schools and curricula.

Table 7-3.   Differentiation Between the College-Bound and Non-College-Bound

Country	Age When Differentiation Occurs	Differentiation Process or Mechanism	Tracks Available Subsequent to Differentiation
E&W	14-16	GCSE examinations	Two tracks: (1) academic study leading to A-levels; (2) vocational courses
France	15	Course grades and brevet examinations	Two tracks: (1) academic lycée; (2) vocational professional lycée
Germany	10-12	Parents and teachers decide	Three tracks: (1) Gymnasium, leading to university study; (2) Realschule, leading to advanced technical, professional, vocational training; (3) Hauptschule, leading to the workforce
Israel	14-15	Parents, teachers, and students decide	Two tracks: (1) academic, including examinations; (2) vocational, with or without examinations.
Japan	15	High school entrance examinations and teacher recommendations	Same course of study for all students, but students attend different high schools based on performance and teacher recommendations
Sweden	14-15	Student preference and course grades	25 different "lines" of study, some vocational, others academic
US	Ongoing, from elementary through high school	Parents, teachers, and students decide	Informal tracking — different courses of study and separate groupings of students exist within individual schools

In the United States, no formal differentiation system exists in elementary and secondary schools. With the exception of certain physically or mentally challenged or disruptive students, all children attend comprehensive schools. However, grouping students by performance or ability is commonly practiced within schools at all levels, including elementary. By upper secondary school, many students are precluded from selecting advanced courses because of their inadequate elementary and lower secondary school preparation.

Programs of Study Leading to Examinations

The caliber of the non-US examinations represented in this book and the corresponding pass rates may raise the question of whether students in these countries are becoming proficient in some subjects at the expense of others. Are students who spend their time studying for science or mathematics examinations forced to neglect other important subject areas? Table 7-4 indicates that whether sets of examinations for college-bound students are specialized or broad-based varies by country.

In all countries, the curricula ensure that students are exposed to challenging courses in core subject areas, such as language/literature, mathematics, science, and history. In England and Wales, however, collegebound students begin to specialize and narrow their focus upon entering the upper levels (last two years) of secondary schooling.

In England and Wales, those studying for their A-levels normally limit their studies to only three subject areas in which they have chosen to be examined.[3] Kelly (1994) reports that students may spend up to 30 percent of their in-school time on subjects outside their examination preparation. Even so, most or all of their homework is devoted to the subjects on which they will be assessed. Universities and departments within universities have varied criteria on the number and subjects of examinations that must be passed by applicants.

Until 1994, French students chose among 38 different baccalauréat tracks, 8 of them being academic courses of study. Each track has its own set of between 7 and 10 courses and examinations. Beginning in 1995, the academic baccalauréats have been reduced to three alternative tracks (Columb, 1995). Students in all tracks take courses and examinations in French literature, philosophy, history/geography, foreign language, and mathematics.

---

[3] In contrast with traditional views of the Advanced-level examinations, an alternative view that considers ages 16 to 18 too young for specialization has had broad support for many years, making reform of the system a continuing subject of political controversy.

Table 7-4.   Number and Range of Subject-Specific Examinations Required for College Entrance

Country	No. of Subject Examinations Taken	Subjects Tested
E&W	3	Universities have their own requirements both in terms of the number of A-level examinations applicants must take, the grades required, and the subject areas covered; subject areas vary widely.
France	7-8	Students take a set of examinations prescribed by their particular baccalauréat track—e.g., students in one popular track take examinations in biology, French, history/geography, mathematics, philosophy, physics/chemistry, and a foreign language.
Germany	4	Students choose four subjects in which to take examinations, three of which must cover (1) language, literature, and the arts; (2) social science; and (3) mathematics, sciences and technology.
Israel	5-11	Students must take a series of examinations worth 2-3 or 4-5 points; half of their points must be in Hebrew, English, mathematics, and the Bible.
Japan	3-4	Individual programs/departments at each university set their own requirements; students generally take at least one examination in the discipline or field to which they apply.
Sweden	6-7	Examinations vary, depending on which of the "lines" students take.
US	—	There are no requirements for taking subject-specific examinations: A small number of students take AP examinations or Achievement Examinations from The College Board on an optional basis; they may take as many or as few as they wish.

In Germany, all Gymnasium students are required to take certain core courses all the way through secondary school, including their final two years as they study for the Abitur examinations. Students eventually choose four subjects in which to take the examinations. At least one examination must be in each of three major curricular areas: language, literature, and the arts; social sciences; and mathematics, science, and technology.

US students who take AP examinations may do so in any subject in which the examinations are offered. There are no government or university requirements in terms of subject areas or numbers of examinations to be taken. The AP examinations, therefore, are not the focus for US students' formal courses of study, which generally include a broad range of subjects.

Characteristics of some of the countries' examination systems take a slight edge off the high stakes associated with taking these examinations. England and Wales, France, and Japan all offer students a "second chance"—that is, students may study for an additional year (or even more) and retake the examinations. This practice is most common among Japanese students, especially those seeking entrance to the most prestigious universities. Also, it is important to note that avenues do exist for students at the end of secondary school in England and Wales (Masters, 1994) and in Japan to enter college without taking the types of examinations discussed in this volume. Again, this situation is most prevalent in Japan.[4] Germans who didn't take examinations at the end of schooling but instead entered the workforce occasionally attempt examinations to secure the Abitur later (Führ, 1989). Sweden has an alternative examination that may be used by students returning to education after having been in the workforce (Gisselberg, 1992).

Though the countries' secondary school curricula generally are well-aligned with their examinations, there are some problems in the correspondence between university curricula and secondary preparation and examinations. Competitive pressures make students gravitate toward examinations that have more prestige for university admission. As a result, some subjects studied in high school in preparation for favored examinations may not correspond to what students intend to study in college. Conversely, subjects that will be needed in college may receive too little emphasis in high school. For example, many students enrolled in biology programs in Japanese universities have studied little biology in high school because examinations in this subject are not among the most prestigious entrance examinations (Nakayama, 1994). In France, the C track, which concentrates on mathematics and physics/chemistry, is the most prestigious track; consequently, the best students take it—regardless of what subjects they plan to take in university (Columb, 1995).

---

[4] At Tokyo University, which is one of the most prestigious schools in Japan, 40 percent of entrants in 1992 were students who had taken the examinations a second time. In general, such students constitute from a quarter to a third of entrants to most universities (Shimahara, 1992; Eckstein and Noah, 1993). Up to 20 percent of entrants to Japanese universities gain entry by teacher recommendations, but not all universities offer this option (Stevenson, 1994), and it is less prevalent in the more prestigious institutions (Ryu, 1992).

Firm information on how students are prepared is hard to uncover. What exactly do teachers typically do to prepare students for examinations? In what specific ways do examinations and national syllabi influence daily instruction? What are the general understandings among a country's teachers about the likelihood of particular topics appearing on examinations? Additional study of these questions is needed.

Since a majority of students who take them pass these examinations, a good look at their content suggests much about what students must be learning. But even seemingly sophisticated examination questions can in reality be less cognitively demanding for examinees if there is explicit coaching of students with similar questions during their schooling. The difficulty can be further diluted if examination topics are stable from year to year. In France, for example, there are indications that students frequently are coached in their classrooms on how to pass the examinations, and that general examination topics are well-known (Columb, 1995: Langlois, 1992).

A discussion on preparing for examinations would be incomplete without noting some activities beyond the formal educational system. In Japan, preparing students for examinations is a major industry, consisting of private "cram" schools (juku) and many commercial study guides with old examinations. In 1987, juku represented a $5-billion-a-year industry—and growing (Leestma et al., 1987). Attendance in juku peaks at 47 percent in ninth grade. While precise data after ninth grade are not available, attendance is somewhat lower, because 30 percent of students are in vocational programs. In France, at least three commercial publishers sell collections of old examinations (e.g., Nathan, Hatier, and Vuibert). In England and Wales and the United States, the bodies that create the examinations sell copies of old examinations.

Creation and Scoring of Examinations

Responsibility for creating and grading examinations lies with teachers and/or university faculty, with the relative role of these two groups varying among countries. In Germany, secondary school teachers write and grade the examinations for their own students; in Japan, university faculty control the entrance examinations under a heavy veil of secrecy. In the other countries, examinations are created by committees of both secondary school teachers and university faculty. As for grading the examinations, in Sweden, teachers grade their own students' work; in Israel, teachers grade other students' work; and in France, England and Wales, and the United States, teachers and university faculty share responsibility for grading. The preparation of the examination usually includes preparation of a scoring guide to standardize scoring.

An important consideration in creating the examination is the emphasis placed on piloting, or field testing, questions for validity and reliability. In the United States, multiple-choice questions are repeatedly tested with first-year college students, and some identical questions are included from year to year to track trends in performance (AP Program, 1992). By contrast, the Educational Testing Service does *not* field test any of the free-response items that makeup 40 to 55 percent of the final examination score. Instead, these questions are written and reviewed by university subject specialists and secondary school teachers. This process is similar to the one used in all the other countries except for Sweden and most of the German Länder. In Sweden, entire sample examinations are tried out on secondary students the year before they are used. In Germany, teachers submit questions for review by the state education ministry or a regional inspector.

System and Student Costs

Funding for creating, administering, and grading examinations tends to be provided by the state, except in Japan and the United States. It is not clear how much these activities cost (Madaus and Kellaghan, 1991). One of the difficulties in determining actual costs is the role played by teachers. French and German teachers play a large role in the development and grading as part of their normal teaching routine, and thus are not paid extra. In the United States, however, teachers are unlikely to assume the added burden of creating and/or grading such examinations without receiving additional pay. In an analogous situation in England and Wales, a pre-A-level assessment was introduced that required extensive additional work by teachers. After a short time, this assessment system unraveled and had to be dramatically reduced in scale, largely because teachers objected to adding a new function, especially without adequate professional development (Black, 1994).[5]

What is clearer is the relationship between the type of items used and the cost of the examinations. Multiple-choice examinations are comparatively inexpensive to grade; while free-response, oral, or practical items become quite costly to grade. Giving students a second chance at the examinations raises costs even higher. Madaus and Kellaghan (1991) have estimated that the cost per student of the English and Welsh General Certificate of

---

[5] A-level examinations are written and graded by selected teachers who are paid for their services. The assessments referred to here were for lower secondary and elementary students.

Secondary Education (GCSE) examinations was $107.[6] They caution, however, that US costs for a similar examination would be higher because of the higher costs of labor.

Only Japan and the United States require students to pay for the examinations directly. For the Japanese universities, the examination fees are a major source of revenue, with the price ranging from $100 per subject for the public universities to an average of nearly $250 for the private universities (Ryu, 1992). Furthermore, students must pay for transportation to the university in order to take the examination (Stevenson, 1994). In the United States, students must pay $67 for each AP examination they take. They must also pay to take the Scholastic Achievement Test or American College Testing Assessment examinations. Some states and school districts make funds available to help defray these costs (AP Program, 1994).

Relevance of Examination Systems to Comparing Examination Difficulty

Which countries' examinations are more difficult than others? Chapter 2 compared the difficulty of examinations by looking at their internal characteristics—length, item types, choice among questions, topic coverage, and performance expectations. No single ordering of countries for difficulty was possible because each examination characteristic led to a different conclusion. (See Table 2-5.) Pressing the issue further, this section looks at some aspects of the countries' examination systems that also must be considered in making judgements about examination difficulty. (See Table 7-5.) Unfortunately, little information was available about some of these aspects, but we nevertheless discuss them to further illuminate the complexity of comparing examinations' difficulty.

How rigorous is the scoring of individual examination questions? Two examinations that are equally difficult for students to answer may not necessarily be scored using the same criteria. At the level of individual questions, how liberally or conservatively are points or partial points awarded for otherwise comparable answers to comparable questions? Since the majority of examinations use free-response items extensively, this issue is quite relevant. While we were able to obtain scoring rubrics for some examinations, they were not available for others.

What are the cutoffs for determining whether the total score allows the student to pass? If cutoffs for two otherwise comparable examinations

---

[6] Comparisons between A-level and GCSE costs are complex. While the individual GCSE examinations are shorter than the A-level examinations, students take more of the former than of the latter.

Table 7-5.   Examination Difficulty: Influence of Examination Systems
For each aspect of difficulty, countries are listed in decreasing order of difficulty.

Features Studied

In how many subjects areas do students typically take examina-tions?	France, Israel, Sweden	approximately 7
	England/Wales, Germany, Japan	3-5
	U.S.	1
How much do examinations count toward determining eligibility for higher education or specific institutions?	England/Wales, France, Japan	examinations largest or sole determinant
	Germany, Sweden, Israel	teacher grades as important
	U.S.	mostly teacher grades

Features Warranting Further Study

How rigorous is scoring of individual questions?

What percentage of the total score must students obtain to pass?

In what ways are daily instructional activities and topic selection related to examinations?

What is the nature and amount of in-school coaching for examinations?

What is the typical out-of-school preparation for examinations?

---

are not very similar, then the examinations' relative difficulties are not the same. Information was available for France and the US, but was less clear in other countries, especially the details of how teacher grades and examination scores are combined. Japan's universities, in particular, have not been willing to release any information about scoring other than the final results, i.e., who is granted admission and who is not.

In what ways are daily instructional activities and topic selection related to examinations? Existing research indicates what courses lead to examinations and the amount of instructional time they are allotted. But more detailed information is sorely needed. Within the available instructional time, what exactly do teachers and students do from day to day to prepare for the examinations? Do students practice with any, some or many previously administered examinations to prepare for the event? Is most of the nature and scope of instruction aligned with examinations, or do teachers go beyond them? For example, do teachers take up topics they feel to be important even if the examinations are unlikely to cover them, e.g., environmental chemistry or applications of mathematics to real-world phenomena? Do teachers make time for laboratory work when examinations do not include it?

What out-of-school preparation for examinations is typical? Some aspects of this are commonly known, e.g., the commercial availability of books containing previous examinations, or the Japanese juku (out-of-school) schools. How much use do students make of the examination books on their own time, or are they poring over them during school hours? Does the level of expectation that parents and the community have for students influence the amount of preparation they do outside of school hours?

How many examinations do students take in addition to the science or mathematics examinations in this study? While AP students in the US typically take one course and its corresponding examination on a single subject, students in England/Wales, Germany and Japan must take from three to five examinations in different subjects. French, Israeli and Swedish students generally must take about seven examinations, one for most academic subjects in their course of study! Overall, facing a large battery of examinations in different subjects is a greater challenge than taking a smaller set of examinations, or just one.

How much do the various curricula concentrate on the subjects in which students will take examinations? The course of study for students in England/Wales mostly is concentrated on the (typically) three subjects for which they will take A-levels, so the instructional time leading to specific examinations in England and Wales is greater than students in other countries receive for similar examinations. While German and Japanese students focus on subjects corresponding to the examinations they will take, students also must maintain studies in other subjects that will not be tested. Students in France, Israel, and Sweden take examinations in most academic subjects they study. US students do not receive any more instructional time for AP courses than for their many other classes.

How much do examinations count toward eligibility for higher education? Students' perceived and experienced difficulty in taking an examination depend in part on what stakes are attached to their performance. Because high school transcripts and SAT or ACT scores have much more bearing on US college admissions than the optional AP examinations, students facing APs are not encountering the same level of pressure as students in other countries where required examinations have decisive influence on eligibility for higher education. In Germany, Sweden and Israel, course grades as well as examination scores determine students' futures while examinations are the largest or sole determinant in England/Wales, France and Japan. Examination scores also can influence the caliber of institution that students can attend or the subject specialization they can study.

The discussions above and those in Chapter 2 show that comparing examination difficulty is a multifaceted endeavor, where the relative difficulties among the examinations indicated by various criteria can be contradictory. Moreover, inadequate information exists about several factors that bear on examination difficulty. Thus, it is important to indicate which criteria have been considered when comparing examination difficulty; otherwise, the comparison could easily lead to inaccurate generalizations.

8

Examination Systems In Seven Countries

Simon Hawkins
Matthew Gandal
Edward Britton

THIS CHAPTER PROVIDES SUBSTANTIAL descriptions of the examination systems of England and Wales, France, Germany, Israel, Japan, Sweden, and the United States. The information presented here is based upon extant literature and original interviews both with experts from the relevant countries and researchers specializing in comparative education. Because the primary purpose of our study was to analyze the examinations themselves, our quest for background information was not exhaustive. During the research, facts routinely surfaced that led to additional questions. When the extensive references consulted did not address these questions, lack of resources prevented our investigating further. Thus, additional research beyond the scope of our current work is warranted.

Due to these limitations, our information is sometimes uneven across countries. Discussions of some system aspects might be more complete for one nation, while comparable information is sparser—or completely lacking—in other countries' descriptions. For example, less information is readily available about Israel and Sweden compared to England and Wales, France, Germany, Japan, and the United States.

England and Wales[1]

Examination Purposes
Some colleges and universities require candidates to have taken and passed two—others, three—Advanced level (A-level) examinations (with the exact requirement influenced by the individual student's course-grades). The better

---

[1] Paul Black, Professor at King's College, University of London, and Gordon Stobart of the London Examinations and Assessment Council gave us detailed comments and suggestions for this section. Additional research assistance was provided

219

the grade on each examination, the better the candidate's chances of being accepted to the school and discipline of his or her choice. The government recently created an alternative route to college entrance, the General National Vocational Qualification Examinations at Advanced Level.[2] Although taken by only about 3 percent of college students in 1994, its use has grown dramatically from 1,000 students in 1993 to 20,000 in 1994 (Masters, 1994).

Number of Examinees, Pass Rates

The government has been trying to increase the number of students who go on to college. In 1988, there were 466,000 full-time undergraduate students; in 1992, there were 700,000 (Masters, 1994). In 1992, 28 percent of the age cohort attended universities (Evans, 1994) and 31 percent of the age cohort took at least one A-level examination (Gandal et al., 1994). Nearly 25 percent of the age cohort earned passing grades in 1992—15 percent passed three or more A-levels, 6 percent passed two, and 4 percent passed one (Stevenson, 1994).

Examination Boards

A-level examinations and their corresponding curricula are developed by seven private examination boards, one for Wales and six for England. While there is no published set of standards or guidelines, the boards have to conform to a common core of content in most subjects. This core is laid down by a recently established national body, the School Curriculum and Assessment Authority (SCAA). SCAA monitors the boards' syllabi and their examinations to ensure consistent and high standards. Appointed by the Secretary of State for Education, SCAA has the power to approve or withhold recognition of a particular examination and curriculum. It also must approve all new examinations and curricula or any significant changes to

---

by C. Coulson, Julie Finch, and V. Nutt of the London Examinations and Assessment Council; Anita Cooper, Pam Jones, and Martin Taylor of the Associated Examining Board; and Janet Bacon of British Information Services.

[2] National Vocational Qualification Examinations are individual vocational/occupational examinations in subjects such as carpentry and car maintenance. General National Vocational Qualification Examinations (GNVQs) have some academic subjects such as science as well as subjects like leisure and tourism. All GNVQs are offered at three levels, with the highest—Advanced—the only one appropriate for college entrance.

existing ones as well as monitor assessment practices and scoring procedures to ensure consistency of standards.

The examination boards were originally set up by universities or groups of universities to help in the admissions process. Although some still have strong ties to universities, they are now independent, nonprofit organizations whose main source of revenue is the fees they charge schools for using their examinations and curricula. Over the years, some boards have come to serve certain regions more consistently, but they have no territorial claims to any particular locality or region. Instead, the boards must attract students and schools with the quality of their materials and their reputations.

The competitive nature of the examination business makes SCAA's role very important. Its seal of approval is essential, and this is what ultimately keeps the boards accountable to the public. Universities exert a considerable amount of influence on the boards, since they—more than any other entity—rely on the boards to give them accurate information about student performance.

Formal Differentiation Between Academic and Nonacademic Students

At age 16, after five years of secondary education, students may choose either to enter the job market or pursue further full- or part-time academic and/or vocational studies. The great majority of pupils take examinations for the General Certificate of Secondary Education (GCSE) in various subjects. Typically, students take GCSE examinations in eight subjects. Those intending to continue their academic studies are expected to earn grades of A, B, or C on an A to G scale in five or more subjects.

In 1991, 62 percent of 16-year-olds chose to continue in full-time academic or vocational education. Another 20 percent chose to engage in part-time study. In 1990, 30 percent of the age cohort went on to study for A-levels (Dobson, 1992).

Programs of Study Leading to Examinations

The national curriculum applies to all primary and secondary school children through the age of 16. The subjects covered include English (and Welsh in almost all schools in Wales), mathematics, science, history, geography, design and technology, music, art, physical education, and—for secondary school students—foreign languages. Up to 70 percent of teaching time is supposed to be spent on these subjects. At four "key stages"—ages 7, 11, 14, and 16—national assessments are conducted in these subjects. Beyond age 14, only English, mathematics, and science are required as full subjects.

Design and technology and a modern foreign language must also be studied as at least half subjects.

After taking GCSE examinations at age 16, students continuing their academic schooling spend two years preparing for their A-level examinations. In comparison with their first five years of secondary schooling (grades 6 to 10)—during which they receive a fairly well-rounded curriculum, including at least the subjects mandated in the national curriculum—students working toward their A-levels usually concentrate on just three subjects; they spend approximately equal time on each. Even if a student chooses to take three A-level examinations in similar subjects, schools generally require some additional study in an area outside the student's selected specialization (for instance, a humanities course for a mathematics/science student, or a general science course for a humanities specialist). Although these additional courses have no homework, they may take up to 30 percent of the class time (Kelly, 1994). At the end of the second year, normally age 18, students take the A-level examinations in each subject they have studied. Students may choose to take one or more examinations earlier or may put off taking an examination.

Students are free to choose the subjects they wish to study. While students have traditionally chosen all three subjects in the same area, such as mathematics/science or the humanities, thereby providing a focus for further specialization at a university, there has been an increasing trend toward studying for A-level examinations in more than one curricular area.

In an effort to allow A-level students to study a wider range of subjects, advanced supplementary (AS) courses were introduced in 1987, and AS examinations in 1989. AS-level courses require half the class time and cover half the subject matter of the A-level. The new option has not yet been widely used by students. In 1992, students took only 53,000 AS subject examinations and over 700,000 A-level examinations (Gandal et al., 1994).

Creation of Examinations

The examination boards hire secondary school and university-level educators on a part-time basis to develop the examinations and the scoring guides that are used by the teachers who grade them. One member of the panel is designated the chief examiner, and he or she oversees the entire process. Separate panels are created for each subject area; business representatives are included on the panels when appropriate. The examination questions must cover the content laid out in the syllabi previously approved by SCAA. Any significant changes in the syllabi or in the makeup of the examinations must be resubmitted for SCAA approval.

The examination panels first meet to discuss the creation of an examination two years before it is given. In this meeting, they analyze the results of the examination just given and review draft questions that they have prepared for the examination to be given in two years. Some questions are rejected; others may be radically amended. The panels meet again about two to three months later to look at second drafts of the new examinations. They check these examinations to ensure that they meet the detailed syllabus objectives; further changes may result from this review. Later, two independent reviewers, who are teachers of the course for which the specific examination is being developed, are involved in the review process. Their task is to ensure that the questions are fair and clear. There are further rounds of review by mail, followed by one last panel meeting to make a final check of the examination (Dobson, 1992).

Scoring and Grading

Secondary teachers in the appropriate subjects score the examinations using the scoring guides developed by the various panels. The examination boards then tabulate these raw scores, and the chief examiner and his or her assistants use that data and their own impressions of student performance to set the raw score cutoffs for each grade. In setting these cutoffs, they take into account those set in previous years to ensure that the grading standards are roughly equivalent. They then rescore the examinations of all students within 1 or 2 points of any of the cutoffs. They also do a statistical analysis of the original scores given so as to identify any graders who are too severe or too generous. If any are identified, the board may enter a correction formula to remedy this. There is also a system for randomly sampling the work of scorers to complement statistical data and to check consistency.

A-level science examinations from some boards include a component for practical work, which is based on a teacher's assessment of a pupil's work done during the course. The teacher has to follow set procedures, and the assessment may be checked by inspection visits or by submission of samples of assessed works.

System and Student Costs

Boards charge fees that are usually paid by the schools out of public funds. Because the boards depend on their examination revenue for survival, they will compete for candidates. One examination board was forced out of business in 1990 because it could not attract enough candidates (Dobson, 1992).

France[3]

Examination Purposes

Students in specialized or general academic lycées work toward a diploma called the baccalauréat, which is awarded based on their performance on a set of examinations taken during their final year. Since its introduction in the Napoleonic era, the baccalauréat traditionally has been targeted at the highest achieving French students, serving as their ticket to higher education and the most prestigious careers.

It is generally accepted that without passing the baccalauréat, jobs with a satisfying career path are hard to find. The baccalauréat course of study should not, however, be seen as strictly a college track. Indeed, of students passing the baccalauréat in 1992, only 58 percent went on to study in a university (Stevenson, 1994).

The most prestigious and selective universities (grandes écoles) require students to pass another examination—the concourse (Langlois, 1992). This examination normally requires one or two additional years of intensive preparation after the baccalauréat. Admission to grandes écoles is extremely competitive, and even a strong performance on the concourse may not ensure admittance. In addition, for some disciplines (medicine, dentistry, and some sciences) in the more popular universities, minimum grades on the examination are required for admission. Nearly 30 percent of 19- to 21-year-olds attend universities, grandes écoles, or classes preparing them to enter grandes écoles (Stevenson, 1994).

Number of Examinees, Pass Rates

The baccalauréat process has evolved considerably over the last decades in response to the government's desire to make it accessible to a larger, more

[3] Thanks go to Jean-Pierre Valentin of the Fédération de l'Education Nationale and Claude Ritzenthaler of the Syndicat des Enseignants-Fédération de l'Education Nationale for helping provide French education statistics. The following individuals also provided information about the brevet and baccalauréat examinations, as well as the French education system in general: Emilie Barrier and Josette LeCoq of the Centre International d'Etudes Pedagogiques, Ministere de l'Education Nationale et de la Culture; Antoine Bodin of the Institut de Reserche sur l'Enseignment des Mathematiques, Université de Franche Comté; Alain Bouvier of the Institut Universitaire de Formacion des Maitres; Françoise Langlois of the Institut Physique Nucleaire de Lyons, University Claude Bernard. John Curack, Desmond Dutcher, Bernard Fronsacq, and Brigitte Pierre of the Embassy of France also assisted us.

diverse population. Most notable, there are now two types of baccalauréat, vocational (divided into professional and technical) and general (academic). Students who receive the vocational baccalauréat may enter the workforce or go on to higher education with a professional focus. Before 1950, only 5 percent of the age cohort typically earned the baccalauréat each year (Embassy of France, 1991). In 1990, 54 percent of the age cohort received a general (academic) or vocational baccalauréat diploma (Langlois, 1992). In 1992, 43 percent of the cohort tried for a general baccalauréat, and 36 percent earned one (Bodin, 1994). The government of France has a declared goal that 80 percent of students will be enrolled in classes leading to a baccalauréat—either academic or vocational—by the year 2000 (Nolan, 1994a).

Regional Examinations

The Ministry of National Education has authority over the content of the curriculum and assessments in primary and secondary schools, but leaves the local administrative responsibilities to 28 regional academies (26 in France, 2 overseas). The French academies are divided into four regional groups. Each group writes its set of examinations under supervision by a representative of the ministry. Consequently, students in all regions of France follow a common core curriculum and are tested on the same knowledge and skills. There is, however, some regional variation. Additionally, some have suggested that standards are not the same across all academies (Eckstein and Noah, 1993).

Formal Differentiation Between Academic and Nonacademic Students

Upon completing the compulsory lower secondary school (collège) at age 16, students may move on to the lycée. They then must decide whether they will enter a three-year stream leading to the baccalauréat diploma—and subsequently, to either higher education or a profession—or a two-year vocational/technical stream leading to the Certificat d'Aptitude Professionelle. Over 75 percent of 17-year-olds and over 55 percent of 18-year-olds study full time in lycées (Embassy of France, 1991).

Students must pass the brevet in order to proceed to the lycée. The brevet has two parts: a series of teacher-designed assessments of student work in 11 subjects during the last two years of collège, and a set of examinations given at the end of collège in French, history/geography, and mathematics. About 85 percent of the age cohort take the examination, and about 75 percent of that group passes it (Gandal et al., 1995). Nolan (1994), in describing the mathematics brevet, reports that all examination items require students to construct responses. Thus, like the baccalauréat, the brevet examination is composed entirely of free-response questions. Just as there are general and vocational baccalauréats, so too are there general and vocational lycées. A committee at each collège makes recommendations as to which lycée each student should attend. Students usually follow this recommendation (Gandal et al., 1995).

Programs of Study Leading to Examinations

During the first year of lycée, students follow a common curriculum that strongly emphasizes the core disciplines—foreign language, French, history and geography, mathematics, natural sciences, physics and chemistry, and sports—complemented by an array of elective course options. The first year is designed to give students a taste of each subject area and prepare them to make important academic and professional decisions in years two and three. Many students repeat a grade sometime during their schooling. In 1992, of the 12th grade students, 32 percent were 19 or older (Stevenson, 1994).[4] Until 1994, French students chose among 38 different baccalauréat tracks— 8 general and 30 vocational. Each general track has its own set of about 7 to 10 courses and corresponding examinations. These eight tracks usually are referred to by their corresponding letter assignments.

Students in all of the general tracks take courses and examinations in some or all of the core subjects previously mentioned. The C track generally was regarded as the most demanding and prestigious, therefore attracting the best students regardless of what they planned to study in university (Colomb, 1995). Some students consequently took more mathematics and physical sciences work than they needed for their intended area of university study (Colomb, 1995). During their final year of secondary school, students in the C track concentrated on science and mathematics. Nearly a third of that year's course work was in mathematics, and nearly a quarter was in science; the remaining course work was in philosophy, history and geography, a foreign language, and physical education. Over their five years of secondary school (lower and upper), C track students devoted 19 percent of their time to science and 21 percent to mathematics. In addition to the philosophy, history and geography, and physical education, they also took extensive course work in French language and literature and less extensive work in a foreign language, art, and economics.

---

[4] It is not uncommon for students who do not pass their baccalauréat on the first try to repeat their final year of secondary school. These students are referred to as "redoublers."

In 1995, the organization of the baccalauréat examinations was changed. The general baccalauréat was reduced to three main tracks (Colomb, 1995):

- L—literary (formerly called A tracks)
- S—scientific (formerly called C, D, D', and E tracks)
- ES—economic and social sciences (formerly called the B track)

At the end of the third and final year of lycée, students take the baccalauréat examinations in the subject areas required by their particular tracks. (The single exception is the French examination, which all students take at the end of year two.) The examinations consist of written and sometimes oral sections, with the written examinations being held on up to four days and the total examination time averaging up to 25 hours. Students are examined in certain core subjects, even if these are not directly related to their area of focus. Students in all general tracks take the same examination in French literature and history/geography; for almost all other subjects, students take examinations that differ by track. Students must take two oral examinations in subjects for which they also take written examinations (Eckstein and Noah, 1993).

Creation of Examinations

Although some administrative authority is given to the 28 academies, the Ministry of National Education plays the dominant role in developing and administering the baccalauréat. For example, it determines the topics to be covered in each year's examinations, as well as the dates and administrative procedures for the examinations. François Langlois (1992) reports the following process:

The 26 academies in France are grouped into four clusters (the two remaining academies represent overseas lycées and form their own cluster), each of which develops its own set of examinations. Each cluster is headed by a rector (chief administrative officer) who appoints lycée and university subject specialists to serve on committees charged with developing the examination questions. All questions must adhere to Ministry of National Education guidelines and grading criteria. The committees are assisted by specialists from inspectorates, local arms of the ministry that oversee the examination process for each cluster. The inspectorates ensure that the Ministry of Education retains final authority over baccalauréat examination standards.

Each committee consists of three or four high school teachers who teach courses in the subject to be examined, and is coordinated by a university professor who has taught for a minimum of six years. The teachers do not receive extra pay for this work, but perform it as an added responsibility. In addition to the regular examination, they prepare a makeup examination for students unable to take the regular one. They also prepare alternative examinations in case there are difficulties with the primary examinations.

The first drafts are written during the holidays (July) by an additional four or five teachers who are asked by the administration to write a draft examination following the general instructions of the baccalauréat. Each of these teachers independently writes a complete examination. The team uses these drafts to create the final examination. The team can choose all of one proposed examination, mix questions from several drafts, or, if necessary, write a completely new examination. It also proposes a scheme for scoring the examination.

Each of the draft examinations is then field tested by one or two teachers, who take the examination under the same conditions as would students. They report on how well they expect their students would do on the examination. The draft examinations, the solutions, and all the reports from teachers are submitted to the inspector. The inspector then verifies that the questions are not too difficult or outside the program and writes a report, with recommendations, to the team. The team prepares a final draft of the examinations, which are, like the previous drafts, subsequently field tested.

All of the final drafts, along with their reports, are given to the rector, who makes the final choice as to which is the main examination, which the makeup, and which the alternative examinations. Conversely, if none are acceptable, the rector can call for a meeting of all those consulted and, if no agreement is reached, select an examination that was accepted—but not used—in a previous year.

Scoring and Grading

A student's final baccalauréat score is determined by taking the composite of the scores for all of the examinations after applying the weights appropriate for each track. For example, in the C track the French literature oral examination receives normal weighting; the written French literature, philosophy, economics, and biology/geology examinations receive a double weighting; the oral foreign language examination is tripled; and the mathematics and physics/chemistry examinations receive five times the normal weight. The grading scale runs from 0 to 20, with 20 being a near-perfect score. A total score of 10 or above is considered passing. Only students who pass may receive a baccalauréat diploma. The diploma also indicates how well the student did:

- 16 or above — Very good
- 14-15.9 — Good
- 12-13.9 — Fairly good
- 10-11.9 — Satisfactory

Those scoring between 8 and 10 are required to complete a second round of examinations, consisting of additional oral examinations in two core subjects. If a student scores higher on the second round of oral examinations than on the first round of written examinations, the regional academy may replace the lower scores with the higher and compute a new overall score. If the overall average is boosted to 10 or above, the student may be granted the baccalauréat diploma. Those whose original score is below 8 fail the examination and may repeat the final year if they so choose (Langlois, 1992).

Baccalauréat examinations are graded regionally by teams of teachers organized by each cluster of academies. For individual cities, the rectors also appoint local juries comprised of university professors and lycée teachers. The juries' responsibilities include administering written and oral examinations, resolving discrepancies between lycée and examination grades, and determining the final results for each candidate. These juries may not include more than one teacher from the same school; teachers may not grade their own students; and teachers may not serve on the jury in the same city for more than one year (Eckstein and Noah, 1993).

System and Student Costs

We do not have specific data on the cost of the baccalauréat. Colomb (1995) reports, however, that French officials are finding it increasingly difficult to offer the baccalauréat for all candidates at the same time and under the same conditions, given the increase in students taking the examination.

Germany[5]

Examination Purposes

While there is no single set of national standards in Germany, there is an earned certificate called the Abitur, whose requirements are quite consistent across the German states (Länder). At one time, passing the Abitur was

---

[5] In describing the German system, we received considerable assistance from Wolfgang Böttcher and his colleagues at the Gewerkshaft Erziehung und Wissenshaft. Additional research assistance was provided by Horst Ewert, Wilfried Krug, and Goetz Reimann of the Embassy of the Federal Republic of Germany.

enough to guarantee a student a place in a university, studying the subjects of his or her choice. Due in part to an increasing number of Gymnasia, however, the number of students gaining an Abitur in Germany has grown beyond the capacity of the university system. From 1960 to 1986, the figure for students obtaining the Abitur grew from 57,000 to 206,000 (Führ, 1989). Consequently, a pass mark on the Abitur is still required for university entrance, but it no longer guarantees students a choice of disciplines. Actual scores on the Abitur have become very important, and entrance into the more popular disciplines, such as medicine, is restricted by quota and often requires additional testing and interviews. In many cases, lengthy waiting lists exist. Some Abitur holders decide not to pursue academics and instead choose apprenticeships or technical training (Eckstein and Noah, 1993).

Number of Examinees, Pass Rates

Only 6 percent of the age cohort in 1960 received the Abitur (Halls, 1994). In 1991, 37 percent of the age cohort took the Abitur, and 35 percent of the age cohort received the Abitur (Stevenson, 1994; Gandal et al., 1995). Typically, 85 percent of those who pass enroll in a university within two years; some of the remaining students are expected to enroll within a few more years (Stevenson, 1994).

State Examinations

In Germany, local control of education is an important tradition. Schooling is primarily the responsibility of each of the 16 German Länder, with the national government playing a coordinating role. For each subject, the specific content of the Abitur examinations and the syllabi for the courses leading up to the examinations are determined by the education ministry in each individual Land. There is, however, a certain degree of uniformity across the Länder. For example, there are common policies for entry to the Abitur, for the number and distribution of subjects in which students must be examined, and for grading. These policies are developed by consensus in the Standing Committee of the Ministers of Education, a body composed of the education ministers from each of the Länder (Führ, 1989). Thus, although curricular differences exist across Länder, the national guidelines effectively maintain a high degree of uniformity.

Formal Differentiation Between Academic and Nonacademic Students

Compulsory schooling begins at age six and usually lasts 9 or 10 years. Children attend Grundschule, or primary school, for four years (six years in two Länder), after which they move into one of three secondary tracks:

- the Hauptschule, the most basic level, lasting through the ninth year of schooling and preparing students to enter employment and receive additional training

- the Realschule, a more advanced level that extends through the 10th year of schooling and prepares students primarily for middle-level, nonprofessional careers (while also allowing access to upper secondary education and subsequent university entrance)

- the Gymnasium, the most academically rigorous secondary school path, aimed at those students interested in attending a university

Some Länder have comprehensive schools from grades 5 to 10 rather than a three-tiered system.

In grades 5 to 10 of the Gymnasium, students take compulsory classes in a wide range of subjects. At the end of the 10th grade, students may qualify for the upper level Gymnasium, covering grades 11 to 13. After grade 10, a substantial number of students leave the Gymnasium, but continue their education until age 18 by combining academic work with full or part-time on-the-job apprenticeships.

Usually, students in Germany are tracked or grouped by ability after the fourth year of schooling (age 10). In some Länder, however, children attend a two-year orientation school, allowing them to postpone tracking until age 12. Parents make the tracking decision in some Länder. In others, school personnel make the decision based on teacher recommendations, student grades, and examination scores (Krusemark and Forsaith, 1995). Exactly which secondary school path students take is determined by their performance in the Grundschule. In 1992, approximately 32 percent of German students were in Hauptschule, 27 percent in Realschule, 28 percent in Gymnasium, and 13 percent in comprehensive schools (Stevenson, 1994).

Programs of Study Leading to Examinations

During the 11th through 13th years of Gymnasium, students receive compulsory instruction in core subject areas;[6] they have elective course options as well. Each subject is taught at both a basic and an advanced level, the latter

---

[6] The reunification of Germany has put stress on the examination system, because the former German Democratic Republic required 12 rather than 13 years of schooling.

involving more rigorous content and longer classes. While Gymnasium have different emphases, most students study German, at least two foreign languages, history, geography, mathematics, physics, chemistry, biology, art, music, physical education, and civics (Krusemark and Forsaith, 1995).

Students begin working toward the Abitur at the beginning of the 12th year. Over the next two years, students must take a total of 28 courses, 22 at the basic level and 6 at the advanced level. Students choose four subjects in which to take the examinations, but are required to take at least one in each of three major curricular areas: language, literature, and the arts; social sciences; and mathematics, science, and technology.

Students usually take the Abitur examination at the end of their final Gymnasium year. Three examinations are written, and the fourth is an oral examination. Examinations in some subjects, such as art and music, may involve performance demonstrations. When a discrepancy exists between a student's course grades and his or her examination scores, additional oral examinations may be given.

Creation of Examinations

In a majority of Länder, teachers are responsible for developing the Abitur examinations that will be given to their students as a responsibility of the job. They create a list of possible examination questions relating to their particular subjects at the beginning of each academic year. Each school then sends its questions to the state education ministry or, in the larger Länder, to a regional "school inspectorate." The questions are each evaluated based on a variety of criteria and either approved or returned to the teachers for improvement (Eckstein and Noah, 1993).

In at least two Länder, including Baden-Württemberg and Bavaria, the Abitur examinations are created and graded at the state ministry level. For every subject, students in these Länder, regardless of individual school attended, take the same examination (Eckstein and Noah, 1993).

Scoring and Grading

The Abitur certificate is awarded based on a combination of students' grades over their final two years of course work and their scores on the examinations. Out of 840 possible points, 540 are reflective of course work (330 from the 22 basic courses, 210 from the 6 advanced); another 300 reflect examination scores. A total score of 280 is considered passing.

In most Länder, teachers both give and score their students' examinations. In fact, part of their preservice training deals with creating and scoring Abitur examinations. Most Länder have a system in place for cross-checking teacher scoring, but the system clearly requires a significant amount of trust to be vested in teachers. For example, in Nordhein-Westfalen, one teacher scores the examinations; the scores are checked by a second teacher. If they disagree, a third teacher adjudicates. The oral examinations are supervised by a team of four, consisting of a chairperson (a senior teacher), an examiner (usually the student's teacher), a recorder, and an observer from another school. The committee votes on the grade (Eckstein and Noah, 1993).

System and Student Costs

We have no information regarding the costs of German examinations.

Israel[7]

Examination Purposes

The Israeli examination system is relatively new, having been reformed in the mid-1970s to give students some flexibility in the course of study they pursue and in how their performance is assessed. The matriculation certification for secondary school students, the Bagrut, is an average of students' course work and scores on Bagrut examinations. Receiving the Bagrut ensures admittance to a university, although not necessarily in the subject of one's choice. Students are admitted to the more popular subjects based on their Bagrut scores and additional examinations. To avoid fatigue and reduce student anxiety, the examinations in subjects that have several tasks (for example, biology) are spread out over several separate time periods (Tamir, 1993).

Number of Examinees, Pass Rates

In 1992, about one-half of the age cohort took the Bagrut, and 40 percent of the age cohort received the Bagrut—that is, a pass rate of about 80 percent (Tamir, 1995).

Governance in the Examination System

Most of the courses that students take for matriculation follow nationally prescribed syllabi. While the syllabi do provide a broad outline for the cours-

---

[7] Pinchas Tamir provided invaluable assistance in explaining the Israeli examination system to us.

es, they leave a great deal of freedom to the teachers. In biology, for example, teachers select topics for research papers and laboratory investigations. They also design resources for students to use during in-depth investigations of the teacher-selected topics. The national examinations are guided by the syllabi, but are sufficiently flexible to accommodate approaches taken by different teachers. Every subject has a professional subject instruction committee that decides on the syllabus and general examination layout, and helps ensure that examinations truly reflect the curriculum (Tamir, 1993).

Individual schools also can design and teach courses on any topic of interest. The school must submit a course design and syllabus to the Ministry of Education; once that has been approved, the school designs an examination and submits that for approval. Of the 22 points that students must accumulate in order to receive their Bagrut, up to 4 may be received for these separately designed courses.

A national science teaching center has been established to foster innovations in science teaching. The center is an umbrella organization with different higher education institutions housing the different subject groups, i.e., physics, biology, chemistry, earth science, and elementary science. Each group works to ensure continuity in Bagrut examinations. The groups operate independently; consequently, examination formats vary by subject. The groups are funded, but not controlled, by the Ministry of Education.

Formal Differentiation Between Academic and Nonacademic Students

Students are directed to begin a Bagrut or other tracks at age 14 or 15. Except for mathematics and English, there is no tracking before then.

Programs of Study Leading to Examinations

Israeli students have a great deal of flexibility in planning their high school course of study. Students must obtain 22 points in order to receive the Bagrut; points are awarded for successfully completing courses and their accompanying examinations. About half of these points must be in the compulsory subjects of Hebrew, English, mathematics, and the Bible. Students select all other courses. In designing their program, students may select between advanced-level courses (worth 4 or 5 points) and simpler courses (worth 2 or 3 points). About half the students taking the Bagrut take a 4- to 5-point science course and examination. Of those, about half take biology, and the rest take physics or chemistry. Any student pursuing the Bagrut may elect to take the 2- to 3-point course and examination, but must have approval from the school to take the 4- to 5-point course and examination.

Students performing at a very advanced level may elect to replace one external examination with an in-depth individual research project and an independent oral examination.

The differentiation of high and low level courses is made during the final two years of secondary school. Until that time, there is only one level for all Bagrut courses. For example, all students take three periods per week of biology in grade 10. In grades 11 and 12, the low level students continue to study biology three periods per week, while high level students receive six periods per week.

The nature of the courses can be quite different and, students taking the low level course would not be able to pass the high level examination. In physics, only the high level examination includes a laboratory practical. In biology, both levels of the examination require laboratory practicals, but the high level version requires students to design and perform a whole investigation in 2.5 hours while the low level version presents a few simple tasks, each requiring 20 minutes at a different laboratory station. Neither level of the chemistry examination has a laboratory practical. Both levels share three sections of questions, but advanced students must complete an additional section with more sophisticated questions.

Creation of Examinations

The Minister of Education appoints a chief inspector for each subject. These inspectors in turn create examination design committees that create the examinations, following the guidelines from the national science teaching center. The science committees, for example, have two members from science education, two teachers, two scientists, and one other member (such as a teacher educator). Each committee spends a year designing the examination, including trials to ensure the examination's validity. Any major changes in the examination must be approved by the respective subject instruction committee. Following the examination, the questions are published and become part of the curriculum. The questions and results of science examinations are entered into the item bank maintained by the various branches of the Israel Science Teaching Center.

However, there are some differences in examination construction from subject to subject. The biology examination, for example, has a practical portion which is designed by a separate small group. Special care is taken in making sure the experiments work as intended.

Scoring and Grading

Matriculation from secondary school, or Bagrut, is based on a combination of examination scores and course work. Students must accumulate 22 points by achieving satisfactory marks in a series of courses. Each of these courses, whether advanced or simple, has an external examination, and the student's final grade in the course is an average of their score on the examination and their scores in the class. Experienced teachers score the examinations during the summer. They do not score their own students' papers. Each paper is scored by two readers. In cases of disagreement, a third reader arbitrates. The practical biology paper is scored only once, because the detailed scoring rubric makes a second scorer unnecessary.

System and Student Costs

We have no specific costs of Israeli examinations; however, the biology and physics examinations are more costly because they include laboratory practicals.

Japan[8]

Examination Purposes

For Japanese high school students planning to pursue higher education, their entire schooling is seen as preparation for the entrance examinations that they must pass to enter college. Examinations are required for all higher education institutions—universities, colleges, and even community colleges (Stevenson, 1994).

The main requirement for admission to public national and local universities in Japan is to pass a two-tiered examination system. To be eligible to

---

[8] For guidance on the Japanese examinations, special thanks are due Robert Leestma of the US Department of Education. Others who helped provide statistics and materials include Hirozumi Anma of Nikkyoso, the Japanese Teachers Union; Robert August of the US Library of Congress; Hiroshi Azuma of Shirayuri College in Tokyo; William Cummings of the State University of New York at Buffalo; Shoichiro Hatsuoka of the Tokyo Office of Postal, Telegraph, and Telephone International; Shin'Ichiro Horie, Masamichi Kono, and Satomi Matsumae of the Embassy of Japan; Shin-Ying Lee and her colleagues at the Center for Human Growth and Development, University of Michigan; Eizo Nagasaki and Katsuhiko Shimizu of Japan's National Institute for Educational Research; Lois Peak of the US Department of Education; and Ling-Erl Eileen Wu of Menlo College in California.

apply to the most prestigious universities,[9] students must pass the University Entrance Center Examination (UECE), formerly known as the Joint First Stage Achievement Test. The UECE, based on the national curriculum, is the closest thing Japan has to a uniform national college admission examination. It is required by all public national and local universities, as well as by a few private universities, and consists of a series of 12 subject-area examinations. It is taken by about 49 percent of college applicants (Feuer and Fulton, 1994). Students applying to the most prestigious of the public universities are required to take UECEs in up to five subjects, while applicants to other universities often take two or more examinations. The UECE is taken in January and is the first step for a prospective public university applicant.

The next step is to take the separate entrance examinations given by each university to which the student applies. The individual entrance examinations are more reflective of the character of the individual universities and are theoretically designed to identify students who are suitable for the education these universities provide. They are also considered more demanding than the UECE. The nature and content of the examinations vary from school to school, and often from department to department within the same university. A clear hierarchy exists at the university level, with some schools, such as Tokyo University, claiming the highest achieving students. The entrance standards from Tokyo University are often considered the most challenging in the nation.

Competition for college admission is intense in Japan, particularly for the most reputable schools.[10] Applicants to national public universities must limit their choices to two institutions, since entrance examinations are only given twice a year, and these universities give their examinations on the same dates. In earlier years, all national public universities had their examinations on the same day, once a year, preventing any choice on students' part.

Private universities and local public institutions, on the other hand, schedule their examinations on alternate dates; they may even admit some students on the basis of grades and recommendations rather than examination scores. There are two recommendation processes. In the first, presti-

---

[9] National public universities, e.g., Tokyo University, are the most sought-after, prestigious universities. They generally have smaller class sizes and better facilities than private schools, yet have a yearly tuition of about $3,600, which is only a quarter of what most private universities charge (Stevenson, 1994). In the last few years, however, the standing of the better private schools has improved enough to attract even some of the best students, who in the past would have applied only to the top-tier public universities (Shimizu, 1994).

[10] Peter Frost (1992) titles a paper on reforms in the Japanese examination system as "Tinkering With Hell."

gious high schools may have a relationship with certain universities, so that the university may take students based on the school's recommendation. The other process, open to students in any high school, is entry based on teachers' recommendations. About 20 percent of all university admissions are from recommendations, but not all universities permit this option (Stevenson, 1994). The more prestigious universities accept few students through this process: public universities admit only 3 to 4 percent of their students via this route (August, 1992).

Number of Examinees, Pass Rates

In 1992-93, about 52 percent of those graduating from high school applied to college (or junior college). Sixty-four percent of these applicants were admitted; thus, just over 32 percent of recent high school graduates were accepted into postsecondary education (Stevenson, 1994). Others (about 15 percent of the age group) decided to retake the entrance examinations after studying for at least one year.[11] At the national public universities in 1984, these reapplicants accounted for some 42 percent of new entrants (August, 1992). The proportion of repeat applicants was lower at private institutions, which are generally less prestigious. Overall in 1984, reapplicants accounted for about a quarter of both applicants and entrants (Shimahara, 1992). That percentage has increased regularly, to about a third in 1992 (Eckstein and Noah, 1993).

National and University Examinations

Like France—and, more recently, England and Wales—Japan has a national curriculum that all elementary and secondary schools must follow. The Ministry of Education, Science, and Culture (Monbusho) has complete authority over curriculum content. Although Japan does have a set of national examinations, the UECE, only some universities require students to take them. Each university or department within a university exercises its autonomy by developing its own set of entrance examinations, the content of which generally reflects the national curriculum. Only the universities' individual entrance examinations are required of all students aspiring to higher education.

---

[11] In Japan, students who repeat examinations are called "ronin," which means "wandering masterless samurai."

Formal Differentiation Between Academic and Nonacademic Students

Neither elementary nor lower secondary schools group students by ability level. In fact, they are prohibited from doing so. Though high school (koto-gakko) is not compulsory in Japan, close to 95 percent of the age cohort attend. Over one-third of high school students attend private schools; the rest attend national, regional, and city high schools, all of which are public.

Unlike elementary and lower secondary school, admission to high schools is competitive, based on both grades and achievement examinations. All sets of high school entrance examinations cover Japanese, mathematics, science, history and geography, and English; some examinations may cover other subjects as well.[12] Students' scores are sent only to the high schools to which they have applied and are never released to the students themselves. Each school conducts its own admissions process using these scores.

Japanese high schools develop distinct reputations based on the quality of students they accept and graduate, leading to an informal, but well-known, hierarchy. The most reputable schools attract and admit the brightest students. These schools tend to feed into the best universities, which, in turn, lead to careers in the most desirable fields and businesses. Though applying to high schools is competitive, students are normally guided by their teachers to apply where they will most likely be accepted. The teacher's role in providing guidance is highly valued, and a large majority of students take their teachers' advice. Most students are in fact admitted to the high schools to which they apply. Students who are not admitted to the high school of their choice may retake the high school entrance examinations a year later (Stevenson, 1994).

Programs of Study Leading to Examinations

Academic upper secondary school students (grades 10 to 12) have extensive distribution requirements. In addition to spending 16 percent of their time on mathematics and 12 percent on science, they must also take courses in Japanese language, geography and history, civics, health and physical education, art, foreign language, and home economics. They also spend 2 or more hours a week in club or home-room activities as part of their school curriculum (Monbusho, 1994).

---

[12] Because over 5,000 different examinations are given each year, there has been little international investigation to date of high school examinations. However, the New Standards Project plans to collect at least a few examinations and have them translated for scholarly inspection (Nolan, 1994b).

A host of private companies offer an array of after-school "cramming" courses, called juku, to prepare students for the examinations and to provide them with enrichment courses in music and the arts. Between 50 and 60 percent of seventh, eighth, and ninth graders supplement their normal schooling with juku classes in one to three subjects, as do 40 to 50 percent of fifth and sixth graders. In high school, an even greater number of students attend such courses.

Students who do not get into the university of their choice have the option of trying again the following year, and for as many years after that as they choose. This practice of reapplication is encouraged in many ways, and a large number of students do it. Some even postpone the application process and spend one or more years after high school as ronin, preparing for the next year's entrance examinations by taking privately offered preparatory courses (yobiko). These courses are aimed solely at preparing students for university entrance examinations, reviewing content covered and providing students with examination-taking strategies (August, 1992).

There are instances where the fit between college and high school curricula is less than optimal because of the examination system. Nakayama (1994, p. 48) notes that "the large number of students with no high school biology who are enrolled in biological science departments at the university level is causing a big problem." Science departments require applicants to pass two science examinations, but do not mandate that one of them be in the intended major. Chemistry and physics examinations are considered more popular than biology. Consequently, even science-oriented high school students typically fail to study any more biology than the topics included in the integrated science course taken at the beginning of high school. The actual number and subject of examinations required varies among universities and even among subjects within a university. In general, though, students must take three to five examinations in at least English, mathematics, and a science subject (Ryu, 1992).

Creation of Examinations

Monbusho contains a special agency, the National Center for University Entrance Examinations, which is responsible for developing and administering the annual UECE taken by public university applicants and applicants to a small number of private universities. This center manages all aspects of the UECE examination process, from registering applicants to distributing and grading the completed examinations.

The UECEs themselves are developed by about 20 different subject-area committees, members of which are selected from a list of university faculty members nominated by their deans or presidents. In addition to formulating the questions, these committees ensure that the examinations conform to the national curriculum, are uniform in style and difficulty level, and are consistent across subjects. Representatives from professional organizations of subject-matter teachers meet annually with the UECE committees to review and discuss examination content, difficulty, and format (Eckstein and Noah, 1993).

The second, and more advanced, set of examinations consists of those developed by the individual universities, each of which has its own process for doing so. At Tokyo University, where the examinations discussed in this volume were developed, the entrance examinations were written by subject-matter committees comprised of professors from various departments. These professors ensure that the examination questions fall within the scope of the national curriculum. Thus, in principle, all questions are answerable if the candidates have a thorough command of the national curriculum (Stevenson, 1994). Usually, however, the questions require a considerable level of academic sophistication. An average of five to seven people write the examinations for the public national universities.

The secrecy that surrounds the highly competitive university entrance examinations makes it very difficult to obtain a detailed understanding of the examination development and scoring processes. Universities do not disclose who constructed or scored examinations or how they did so. Faculty take great care to avoid any prospect of inappropriate interactions with prospective students or their high schools; they may not even tell their family members of their role in the examination process.

Scoring and Grading

The Tokyo University examinations are scored by Tokyo University professors.

System and Student Costs

The national universities charge about $100 per subject examination; the private universities charge about $250 per subject examination (Ryu, 1992). Further, students must pay for their travel to the university to take the examinations (Stevenson, 1994). Examination fees are an important source of revenue for Japan's private universities.

Sweden[13]

Examination Purposes

Sweden has no matriculation examination, but students do take Central Tests, national examinations that assess their work in specific courses. Their scores on these examinations account for a significant amount of their final grade in these courses: the exact amount varies by teacher. In addition to affecting individual students' grades, the examinations help standardize grading results across the country. The distribution of a class's final grades must roughly match the distribution of its scores on the national examination. This matching between course grades and examination scores takes place only at the class level, giving teachers the freedom to determine how best to match an individual student's overall performance with his or her performance on the examination. For example, if a student is sick on the day of the examination, the teacher would still be able to evaluate his or her performance for a final course grade.

The examinations are formatted similarly to other examinations given during the school year, making them somewhat familiar to students. Each examination is cumulative, addressing all the content students have covered in the subject, although placing a greater emphasis on material studied more recently. College-bound students take examinations in six or seven subjects, depending on their course of study, or line. To reduce the examination burden, the examinations are spread out over the final two years of schooling, instead of holding all the examinations in one short period (Eckstein and Noah, 1993).

Students receive a grade of from 1 to 5 for each course, with 1 the lowest and 5 the highest. The average of these grades determines their eligibility for university. Different programs within universities require different grade point averages. Forty percent of all university admission slots are set aside for those entering based on their scores on the optional Swedish Scholastic Test, combined with their final marks from upper secondary school and/or work experience. Students who are not able to enter university directly from school may be able to attend at a later date (Gisselberg and Johansson, 1992).

---

[13] Kjell Gisselberg provided invaluable assistance in explaining the Swedish examination system to us.

Number of Examinees, Pass Rates

All high school students pursuing the academic lines (about 45 percent) take the national examinations (Gisselberg and Johansson, 1992). Their scores on these examinations are used by teachers to calibrate course grades, which are in turn used by universities in setting admissions standards.

Governance in the Examination System

The Swedish secondary school curricula are developed by the National Board of Education. Generally, some of the people who helped develop the examination also serve on the board designing the curriculum, so there are definite linkages between the two. Curricula plans are not particularly prescriptive, leaving teachers free to decide how they will teach the various content areas.

Formal Differentiation Between Academic and Nonacademic Students

About 90 percent of students attend the noncompulsory upper secondary school. In upper secondary school, students choose a course of study from among the 25 available lines. These lines range from the vocational to the academic and last from two to four years. There are only a limited number of openings in each of the lines, so the more popular lines must turn some students down, using grades as the criteria. About 45 percent of the age cohort choose to pursue one of the theoretically oriented lines, 7 percent of the age cohort choose natural sciences, and 12 percent choose technology (Gisselberg and Johansson, 1992).

Tracking begins during the 10th year of schooling, when students choose the line they will pursue. Before this point, only mathematics is tracked. In seventh grade, students may choose between general and special mathematics, with most students choosing the more advanced special mathematics. Many schools do not teach these as separate courses, but they individualize instruction for students within a general mathematics course.

Programs of Study Leading to Examinations

In Swedish upper secondary school (generally grades 10 to 12, but sometimes 10 to 13), students follow one of 25 lines, each with a specific vocational or academic focus. Of the 25 options, there are five academic/theoretical lines—natural science, technology, social sciences, economics, and liberal arts. Also, upper secondary school as a whole has a central curriculum. All students must take Swedish, physical education, 30 lessons in environmental knowledge, and civics or orientation for working life; they also take three years of art, social studies, and other courses. In addition, students in all theoretical lines and some vocational lines must study English. Students in the natural science line take three years each of mathematics, physics, and chemistry and two years of biology. They spend 16 percent of their time on mathematics and 25 percent on the sciences (Gisselberg, 1992).

Reforms presently under way will eliminate the central examination system. In the coming years, there will be a criterion-referenced marking system instead of the current norm-referenced system. Central examinations will be given only in English, Swedish, and mathematics. In mathematics, there will be five courses—A, B, C, D, and E. Some courses will be tested once a year; others, twice a year.

Discussions are ongoing about some kind of centralized item banking system in physics and some other subjects. Under this system, schools could ask for a specified examination from the item bank repository at almost any time. These changes are caused by the desire to give schools total freedom to organize and time the different courses as they like.

Creation of Examinations

The Central Tests are linked directly to a national curriculum. The examination writing process is supervised by a committee of experts in the subject appointed by the National Board of Education. Frequently, the board cedes its appointment power to a more decentralized institution. For instance, the Department of Educational Measurements at the University of Umeå appoints the members of both the physics and the chemistry committees. Each committee has six to eight people from a wide range of fields, including teacher educators, scientists, and teachers. About a year and a half before the examination is to be given, the committee invites teachers to write questions for the examination; members then analyze the submitted questions. They select roughly three times the total number of questions needed and field test them with students about three weeks before these students take the current year's examinations (in the hope, not always realized, that they will be motivated to do their best). After analyzing the results of these trials, the committee then assembles the examination.

Scoring and Grading

Each teacher grades his or her own students' examinations following guidelines developed by the appropriate committee. If teachers have any problems or questions during the grading, they may address these to the committee that prepared the examination. Before 1982, upper secondary school inspectors visited schools regularly to spot-check the grading of examinations. There has been little oversight since 1982, however. The committee analyzes a sample of students' raw scores to determine the distribution of scores and develop the desired distribution of final marks for individual students. Each teacher decides how to combine the examination score with his or her own assessments of the students to arrive at the students' final grades. However, general distribution of grades in each class must match the general distribution of examination scores in each class (Gisselberg and Johansson, 1992). For example, if 10 students receive a score of 70 percent on the examination, then approximately 10 students in the class should also receive comparable course grades.

System and Student Costs

Students do not pay for the Central Tests. The cost for developing the examinations is about $150,000 to $200,000 per subject. There are no grading costs, because teachers grade their own students' examinations as part of their regular teaching duties.

United States[14]

Examination Purposes

There are no standard college admission criteria in the United States that are based on student performance. Of the approximately 3,600 institutions of higher education, nearly one-third have no entrance standards at all, operating instead under "open admissions" policies. Those colleges and universities that *do* have selective admissions policies develop their own selection criteria based largely on some combination of course credits; course grades; privately developed and administered examinations; and other factors such as extracurricular activities, student essays, and interviews. The more prestigious schools tend to set high standards.

While there is no single examination that college-bound students must take in the United States, there are several examinations used by admissions offices in making their selections. All of these examinations are developed

[14] In the United States, information on Advanced Placement Examinations, American College Tests, and the Scholastic Achievement Test was provided by Ray Nicosia and Lisa Pagliaro of the Educational Testing Service, Kelley Hayden of the American College Testing Program, and Mary Frances Studenmund of The College Board.

and administered by private organizations. The best known and most widely taken examinations are the Scholastic Achievement Test I (SAT I), developed by the Educational Testing Service (ETS) in cooperation with The College Board; and the American College Testing Assessment (ACT), developed by the American College Testing Program. Neither of these examinations is linked to any particular curriculum taught in the schools.

The College Board and ETS also offer two other series of examinations, SAT II (formerly known as the Achievement Test) and Advanced Placement (AP) examinations, both of which are considered more curriculum-based— and therefore, more indicative of academic achievement—than the SAT I or ACT. SAT II examinations are offered in the subjects of English composition, literature, US history, European history, mathematics, biology, chemistry, physics, and six foreign languages. Advanced Placement examinations are given in art history, studio arts, biology, chemistry, computer science, economics, English (language and literature), French (language and literature), German, government and politics, European history, US history, Latin, mathematics (calculus and a planned statistics examination), music theory, physics, psychology, and Spanish language and literature. Both sets of examinations require greater subject-matter knowledge than the SAT I or ACT. AP examinations are regarded as more advanced than the SAT II in part because they are based on college-level content, while the latter are a measure of high school-level content. Another important difference between the two sets of examinations is that AP teachers are supposed to follow a specific curriculum in preparing students for the examination in each subject.

Although each admitting institution has its own policy for weighing AP and SAT II examinations in its admissions criteria, students with high scores are generally considered to have an advantage over other applicants. Most schools even allow students with high AP scores to earn advanced credit in certain subjects. High schools may offer AP courses in some or all of the subjects mentioned above. While The College Board publishes guidelines for AP courses, including suggested labs, there is no requirement that teachers follow them. In 1993, 48 percent of US high schools offered at least one AP course (AP Program, 1994). AP examinations are *not* a required part of AP courses.

Number of Examinees, Pass Rates

From 1987 to 1992, the number of students taking AP examinations grew by 48 percent (AP Program, 1992). Even so, by 1993 only about 6.6 percent of all 18-year-olds took at least one AP examination; and only about 4.3 percent passed, for a pass rate of 66 percent (AP Program, 1994; NCES, 1993).

Governance in the Examination System

The education system in the United States is built upon a strong tradition of local control. Historically, the federal government has played a limited role in areas such as curriculum and assessment, particularly when compared to such countries as France, Japan, and—since their recent reforms—England and Wales. States and local districts have shared the bulk of the responsibility in these areas, and each has approached the task differently. Some states have assumed a significant leadership role, defining a core curriculum and developing statewide assessments. Others have entrusted districts with these responsibilities. Only a few states, including California and New York, administer examinations that may influence university admission and are tied to the curricula covered in high school; these examinations, however, are all voluntary.

The result of this highly decentralized system has been wide variation in curricula and performance expectations across the country. The ongoing federally sponsored efforts to establish national standards in specific subjects is a historic first attempt to achieve some national consensus on what should be taught. These efforts were first catalyzed by the National Council of Teachers of Mathematics' development of mathematics standards (1989). Currently, the National Academy of Sciences is developing science standards (National Reserach Council, 1994), and the American Association for the Advancement of Science has—through its Project 2061—produced standards it calls benchmarks (AAAS, 1993).

Formal Differentiation Between Academic and Nonacademic Students

Differentiation of students follows no formal system in US schools. All students, except for some disabled or disruptive students, attend comprehensive schools. Within these schools, however, students are frequently grouped based on their ability. While there may be no formal mechanism that prevents students from moving from one group to another, in practice it is frequently difficult to move to a more advanced group. The problem is particularly acute in secondary school. Here students are rarely able to move up, since their earlier educational experiences did not adequately prepare them for more rigorous academic efforts.

Programs of Study Leading to Examinations

There is a specific curriculum that Advanced Placement teachers are supposed to follow in preparing students for the examination in each subject.

The College Board provides participating schools with course outlines in each of the AP subjects offered. These outlines provide teachers with a clear framework for the subject matter that ought to be covered.

Creation of Examinations

The College Board has standing subject-matter committees with rotating membership in each AP examination area. Members are AP high school teachers and college faculty, and are charged with developing examination questions that are well-aligned with existing course descriptions. Both the AP courses and examinations are meant to reflect college-level work. To ensure that the AP materials are keeping pace with university expectations, The College Board conducts curriculum studies every four years (or more frequently if needed) involving up to 200 universities (The College Board, 1994).

Multiple-choice questions are written by secondary school teachers and subject-matter specialists and reviewed by the committee. The questions are then tested on college students who have just completed a comparable college-level course and subsequently, as appropriate, added to a bank of available multiple-choice items. The committee writes the free-response questions itself and does not field test them.

Each fall, the committees meet to discuss the previous year's examination, perform a final review of the examinations to be given in the spring, and begin drafting the free-response section of the next year's examination. As input to this meeting, each committee member drafts a series of possible free-response items. The committee meets again in the spring to further prepare the next year's examinations. They review the free-response section and draft the multiple-choice section by drawing from the question bank. Concurrently, new multiple-choice questions are drafted.

Scoring and Grading

The AP examinations consist of both multiple-choice and free-response sections. The multiple-choice items are all graded by computer. This score is the number of correct answers minus a fraction of the incorrect answers; answers left blank are not included in the scoring. The rest of the questions (essay writing, problem solving, and other free-response items) are graded by specially trained high school teachers and university professors. To ensure the greatest accuracy, each examination is evaluated by four teachers or professors.

The cutoffs between grades are set by the chief reader for each examination, who oversees the various stages of the examination process in consultation with ETS professional staff. To set these boundaries, the chief reader may statistically analyze the score distribution and/or make comparisons with scores from previous years. AP examinations are graded on a scale of 1 to 5. Scores of 3 to 5 are considered passing. In 1993, 66 percent of examinations received passing scores (AP Program, 1994).

System and Student Costs

Students pay to take the examinations—$67 in 1993 (AP Program, 1993a)—and these fees are the major source of revenue for the testing organizations. Some states pay the fee for disadvantaged students. The College Board itself makes some money available for fee reductions for those students who cannot afford its examinations (AP Program, 1994).

Appendix A

Augmented TIMSS Curriculum Frameworks: Topics in the Third International Mathematics and Science Study (TIMSS)

The most detailed categories that follow were prepared by Pinchas Tamir (Biology), Dwaine Eubanks (Chemistry), Kjell Gisselberg (Physics) and John Dossey (Mathematics) for purposes of this examination study; all other categories constitute the curriculum frameworks for TIMSS which are fully described in Robitaille et al., 1993.

This appendix only lists the topics of the Curriculum Frameworks. Some topics common to both chemistry and physics are found in both those sections of the appendix. Aspects of the frameworks not included here are Performance Expectations and Perspectives.

Augmented TIMSS Curriculum Frameworks

SCIENCE

Biology

Diversity, Organization, Structure of Living Things

Plants
Algae
Fungi and mushrooms
Mosses
Ferns
Seed producing plants

Animals
Invertebrates
Unicellar animals
Coelenterates
Worms
Insects
Spiders
Vertebrates
Fishes
Amphibians
Reptiles
Birds
Mammals

Other organisms
Microorganisms
Diversity of microorganisms
Viruses
Roles in recycling
Microorganisms and Man

Organs, tissues
Complementarily between structure and function
Cells
Cell structure and function
Types of cells
Cell reproduction

Life Processes and System Enabling Functions

Life Processes and Systems
Photosynthesis, energy capture, storage and transfer
Respiration, mitochondria
Digestion and excretion
Other energy handling

Sensing and Responding
Biofeedback and homeostasis
Sensory systems, responses to stimuli

Biochemical Processes in Cells
Metabolism, protein synthesis, enzymes
Regulation of cell functions
Cell water relationship

Life Spirals, Genetic Continuity, Diversity

Life Cycles
Life cycles of plants, insects etc.
Reproduction, aging, death
Cell division, differentiation, succession

Reproduction
Reproduction in seed plants
Sexual reproduction
Human reproduction
Vegetative reproduction

Variation and Inheritance
Meiosis
Mendelian genetics
Molecular genetics
Population genetics
Biotechnology and application of genetics

Evolution, Speciation, Diversity
Variation
Evidence of Evolution
Mechanisms of evolution: Lamarckism
Implications of evolution

Interactions of Living Things

Biomes and Ecosystems
Tundra and deserts
Rain forest and wetland other biomes or ecosystems
Habitats and Niches
Habitats and biotopes
Niches, endangered species

Human Biology and Health

Nutrition
Foods, vitamins, minerals etc.
Balanced diets
Diseases and Health
Prevention of disease, maintaining good health
Causes of diseases
Remedies

Biochemistry of Genetics
Structure of DNA
Replication of DNA
Transformation DNA to RNA
Mutation, gene expression
Operon model in bacteria
Implications for society, genetic engineering

Interdependence of Life
Food chains webs
Adaptations to habitat conditions
Competition among organisms
Symbiosis, commensalism, parasitism
Man's impact on the environment

Human Biology
Organ systems, organs, tissues
Cells
Energy handling
Sensing and responding
Life cycle
Reproduction
Genetics

Animal Behavior
Territorialism
Social grouping (beehive, herds)
Mating behavior and selection
Migration of birds, fishes, butterflies
Rearing the young
Learned behavior

Evolution
Biochemistry of genetics
Interdependence of life
Human behavior
Man's impact on environment

Augmented TIMSS Curriculum Frameworks

Chemistry

Matter

Classification of Matter
Elements, compounds, mixtures
Solutions, colloids

Physical Properties
Mass, volume
Density
Physical states

Chemical Properties
Evidence of change
Combination reactions
Decomposition reactions
Addition reactions
Substitution reactions

Structure of Matter

Atoms, Ions, Molecules
Elementary atomic theory
Periodicity, metals, nonmetals
Ionic compounds
Covalent compounds
Formulas, equations, nomenclature
Mole concept

Macromolecules and Crystals
Crystal structure
Bonding in crystalline solids
Polymers

Subatomic Particles
Protons, electrons, neutrons
Isotopes
Quantum objects
Electromagnetic radiation and matter
Quantum numbers, orbital energies
Electron configuration, periodicity

Quantum Theory
Photoelectric effect
Line spectra
Matter waves
Uncertainty principle

Physical Transformations

Physical Changes
Gases
Liquids, solids
Phase changes, phase diagrams
Solutions
Colligative properties

Explanations of Physical Changes
Dynamic equilibrium
Inter-particle forces
Dispersion and flocculation of colloids

Kinetic Theory
K-M view of gases
K-M view of liquids, solids

Chemical Transformation

Chemical Changes
Acids, bases
Ionic reactions

Oxidation, reduction

Coordination chemistry

Explanations of Chemical Changes
Ionization energy, electron affinity, electronegativity
Ionic and covalent bonds
Molecular shape
Periodic trends of reactivity

Rate of Change, Equilibria
Reaction rates, rate laws
Catalysis, activation energy
Reaction mechanisms
Equilibrium expressions

Energy, Chemical Change
Calorimetry
First law of thermodynamics
Second law of thermo-dynamics

Organic and Biochemical Changes
Hydrocarbons
Organic oxygen and nitrogen compounds
Addition and substitution reactions
Mechanisms of organic reactions
Biologically important carbon compounds

Nuclear Chemistry
Alpha and beta particles, gamma rays, and neutrons
Mass defect and nuclear binding energy
n/p ratios and nuclear transformations

Kinetics of nuclear decay
Fission, fusion
Biological effects

Electrochemistry
Electrolysis
Electrochemical cells
Free energy, cell potentials
Practical electrochemistry corrosion

Physics

Mechanics

Physical Properties
Mass and volume
Density
Physical states

Energy Types, Sources, Conversions
Work, energy, power
Kinetic and potential energy
Energy types and trans-ormations
Energy sources

Types of Forces
Gravitation
Friction
Tension
Static equilibrium (incl. simple machines, centre of gravity)
Nuclear forces

Time, Space and Motion
Measurement of space, time and mass
Linear motion
Projectile motion
Circular motion
Motion in two dimensions

Dynamics of Motion
Laws of linear motion
Linear momentum, conservation of momentum and/or energy
Laws of circular motion
Angular momentum, moment of inertia, rotational kinetic energy

Fluid Behaviour
Pressure, Archimedes' principle
Liquid flow, continuity equation
Bernoulli theorem

Electricity & Electromagnetism

Electricity
Electric charge, conductors, insulators, current
Electric field, potential, voltage, resistance
Capacitors, series and parallel, dielectrics
Charging and discharging of capacitors

Electric power and energy

DC circuits
AC circuits
Electronics, semiconductors
Electromagnetic oscillations

Augmented TIMSS Curriculum Frameworks

Electromagnetism & Magnetism
Magnetic forces, magnetic fields
Electromagnetism
Induction
Self inductance
Charges in electric and magnetic fields

Waves, Sound, Light

Wave Phenomena
Simple harmonic motion, pendulums
Transverse waves
Longitudinal waves
Superposition of waves, interference
Doppler effect

Diffraction, the electromagnetic spectrum
Radiowaves, radio transmission
Sound and Vibration
Standing waves
Infrasonic and ultrasonic waves
Intensity of sound

Light
Reflection and refraction
Light intensity, luminosity
Fibre optics
Polarized light

Thermophysics

Heat and Temperature
Heat and energy, changes of state
Thermal expansion
Thermal equilibrium, conduction
Thermoelectricity
Emission and absorption of heat radiation
Physical Changes
Gaseous state
Pressure, volume and temperature relationships
Partial pressures
Diffusion, effusion
Real gases
Properties of liquids
Properties of solids

Crystal structure
Phase changes
Heating and cooling curves
Phase diagrams
Formation of solutions
Solution concentration
Effects of temperature and pressure on solubility
Colligative properties
Explanations of Physical Changes
Freezing and boiling of pure substances
Intermolecular forces
Dynamic equilibrium

Ion-dipole and dipole-dipole forces
Hydrogen bonding
Freezing point depression
Osmosis, dialysis
Colloidal dispersions
Kinetic Theory
K-M view of gases
K-M view of liquids and solids
Energy and Chemical Change
Calorimetry (chemical reactions)
First Law of thermodynamics
Second Law of thermodynamics

Atomic and Quantum Physics

Atoms, Ions, Molecules
Dalton's atomic theory
Atomic masses

Subatomic particles
Nuclear atom
Metals, nonmetals
Periodicity

Ionic compounds
Molecular compounds
Naming compounds
Formulas and equations

Mole concept
Macromolecules, Crystals
Ionic crystals
Network solids
Metallic solids
Organic polymers
Inorganic polymers
Biopolymers
Subatomic particles
Protons electrons and neutrons
Isotopes
Properties of quantum objects
Electromagnetic radiation and matter
Exclusion principle and quantum numbers
Orbital shapes, energies
Multi-electron atoms
Electron configurations
Electron structure and periodicity
Quantum Theory
Unspecified
Photoelectric effect
Line spectra
Matter waves
The uncertainty principle
Quantum effects, tunneling

Nuclear Physics

Alpha and beta particles, gamma rays and neutrons
Mass defect and nuclear binding energy

Relativity and Cosmology

Cosmology
Beyond the solar system
Evolution of the universe

Chemical Changes
Acid-base reactions
Acid-base stoichiometry
Acid-base definitions
Ionic reactions
Combustion reactions

N/p ratios and nuclear transformations
Kinetics of nuclear decay
Fission, fusion

Relativity
Basic postulates of theory
Mass-energy correspondence

Other oxidation-reduction reactions
Oxidation numbers
Balancing redox equations

Biological effects of radiation
Nuclear energy transformations
Nuclear models

Relativistic energy and momentum
Lorentz transformations and addition of velocities
Minkowsky space

Augmented TIMSS Curriculum Frameworks

Science and Other Disciplines

Standard units (customary and metric)
Quotients and products of units
Dimensional analysis
Estimation of measurements and errors of measurements
Precision and accuracy of measurements
Points, lines, segments, half-lines and graphs
Circles and their properties

Concept of vectors
Vector operations (addition and subtraction)
Vector dot and cross product
Slope and gradient in straight line graphs
Trigonometry of right triangles
Representation of relations and functions
Interpretation of function graphs

Logarithmic and exponential equations and their solutions
Uncertainty and probability
Limits and functions
Growth and decay
Differentiation
Integration
Differential equations

Mathematics

Numbers

Whole Numbers

Meaning
Uses of numbers
Place value and numeration
Ordering and comparing numbers

Operations
Addition
Subtraction
Multiplication
Division
Mixed operations

Properties of Operations
Associative properties
Commutative properties
Identity properties
Distributive property
Other number properties

Fractions and Decimals

Common Fractions
Meaning-representation of common fractions
Computations with common fractions and mixed numbers
Decimal Fractions
Meaning-representation of decimals
Computations with decimals

Relationships Between Common and Decimal Fractions
Conversion to equivalent forms
Ordering of fractions and decimals
Percentage
Percent computations
Percentage problems (increase, decrease,....)

Properties of Common and Decimal Fractions
Associative properties
Commutative properties
Identity properties
Inverse properties
Distributive properties
Cancellation properties
Other number properties

Integers, Rational, and Real Numbers

Negative Numbers, Integers, and their Properties
Concept of integers
Operations with integers
Concept of absolute value
Properties of integers

Rational Numbers and their Properties
Concept of rational numbers
Operations with rational numbers
Properties of rational numbers
Equivalence of differing forms of rational numbers
Relation of rational numbers to terminating and recurring decimals

Real Numbers, their Subsets, and their Properties
Concept of real numbers (including concept of irrationals)
Subsets of real numbers (Z, Q, W, N)
Operations with real numbers
Properties of real numbers (density, order, completeness)
Operations with absolute value

Augmented TIMSS Curriculum Frameworks

Other Numbers and Number Concepts

Binary Arithmetic or Other Number Bases
Exponents, Roots, and Radicals
  Integer exponents and their properties
  Rational exponents and their properties
  Roots and radicals and their relation to rational exponents
  Real exponents

Number Theory
  Primes and factorization
  Elementary number theory (primes, lcm, gcf, diophantine problems)
Systematic Counting
  Tree diagrams, listing, and other forms
  Permutations
  Combinations
  Generating functions

Complex Numbers and their Properties
  Concept of complex numbers
  Algebraic form of complex numbers and their properties
  Trigonometric form of complex numbers and their properties
  Relation of algebraic and trigonometric forms of complex numbers-DeMoivre's theorem

Estimation and Number Sense

Estimating Quantity and Size
Rounding and Significant Figures

Exponents and Orders of Magnitudes

Estimating Computations
  Mental arithmetic
  Reasonableness of results

Measurement

Units

Concept of measure (incl. non-standard units)
Standard units (Customary and Metric)

Use of appropriate instruments (ruler, protr.)
Common measures (length, area, volume, capacity, time/calendar, temperature, angles, weight/mass,...)

Quotients and products of units (km/hr. , m/s^2)
Dimensional analysis

Perimeter, Area, Volume, and Angles

Computations, formulas, and properties of length and perimeter
Computations, formulas, and properties of area

Computations, formulas, and properties of surface area
Computations, formulas, and properties of volume

Computations, formulas, and properties of angles

Estimation and Errors

Estimation of measurements and errors of measurement

Precision and accuracy of measurements

Geometry-Form
(Position, Visualization, and Shape)

Two-Dimensional Geometry

Coordinate Geometry
Line and coordinate graphs, midpoints
Equation of lines in the plane
Conic sections and their equations
    Parabola
    Ellipse
    Hyperbola (including asymptotes)

Basics
Points, lines, segments, half-lines, and rays
Angles
Parallelism and perpendicularity
    Parallel postulate
    Perpendicularity
Basic compass/straightedge constructions

Polygons and Circles
Triangles and quadrilaterals: classification and properties
    Triangles, Quadrilaterals
Pythagorean theorem and its applications
Other polygons/ properties
Circles and their properties
Locus problems

Three-Dimensional geometry

3-dimensional shapes and surfaces and their properties
Planes and lines in space

Spatial perception and visualization
Coordinate systems in three dimensions

Equations of Lines, Planes, and Surfaces in Space
Equation of line in space
Equation of plane in space
Equation of quadric surface in space
Equation of a sphere

Vectors

Concept of vectors
Vector operations (addition and subtraction)
Vector dot and cross product

Norm and resolution of vectors
Normal vector to line/plane
Matrix operations

Eigen values/eigen vectors
Vector/matrix form of transformation

Geometry-Relation
(Symmetry, Congruence, and Similarity)

Transformations

Patterns, Tessellations, Friezes, Stencils, etc.

Symmetry
Line symmetry

Reflectional symmetry
Rotational symmetry

Augmented TIMSS Curriculum Frameworks

Transformations
Translations
Reflections
Rotations
Dilations

Compositions of trans-formations
Group structure of trans-formations
Fixed points of transformation

Congruence and Similarity

Congruences
Concept of congruence (segments, angles,....)
Triangles (SSS, SAS, ...)
Quadrilaterals
Polygons
Solids

Similarity
Concept of similarity (proportionality)
Triangles (AA, SSS, SAS, ...)
Quadrilaterals
Polygons
Solids

Constructions using Straightedge and Compass

Proportionality

Proportionality Concepts

Meaning of Ratio and Proportion

Direct and Inverse Proportion
Direct variation
Indirect variation
Other proportional relationships

Proportionality Problems

Solving proportionality equations
Solving practical problems with proportions

Scales (maps and plans) and Rates
Proportions based on similarity

Slope and Trigonometry

Slope and gradient in straight line graphs

Trigonometry of right triangles

Linear Interpolation and Extrapolation

Interpolation

Extrapolation

Functions, Relations, and Equations

Patterns, Relations, and Functions

Number patterns
Relations and their properties
Functions and their properties (range/domain,....)
Representation of relations and functions

Families of functions (graphs and properties)
Operations on functions
Related functions (inverse, exp/log, derivative,....)
Relationship of functions and equations

Interpretation of function graphs
Functions of several variables
Recursion
Hyperbolic functions

Equations and Formulas

Representation of numerical situations
Informal solution of simple equations
Operations with expressions
Equivalent expressions (factorization, simplification, partial
fraction decomposition, and solution tests)
Linear equations and their formal (closed) solutions
Quadratic equations and their formal (closed) solutions
Polynomial equations and their solutions
Trigonometric equations and identities (including law of
cosines and sines)

Logarithmic and exponential equations and their solutions
Solution of equations reducing to quadratics, radical equations,
absolute value equations, ...
Other solution methods for equations (successive ap-
proximations, bisections,...)
Inequalities or their graphical representation

Systems of equations and their solutions
Substitution
Linear combinations/Gauss-Jordan
Matrix solution
Systems of inequalities
Substituting into or rearranging formulas
General equation of the second degree and its interpretation
Parametric equations
Rational equations

Data Representation, Probability, and Statistics

Data representation and analysis

Collecting Data from Experiments and Surveys
Data from experiments
Data from surveys
Representing data
Bar, line, circle, and histographs
Stem-and-leaf, box-and-whisker,....
Scatterplots
Multivariable plots

Interpreting tables, charts, plots, and graphs
Bar, line, circle, and histographs
Stem-and-leaf, box-and-whisker,....
Scatterplots
Multivariable plots
Kinds of scales
Nominal
Ordinal
Interval
Ratio

Measures of central tendency
Mean
Median
Mode
Measures of dispersion
Range
Variance
Standard deviation
Interquartile range

Augmented TIMSS Curriculum Frameworks

Sampling, randomness, and bias
   Types of sampling
   Randomness
   Bias–detection and avoidance
   Variability of sampling mean
   Prediction and inferences from data

Fitting lines and curves to data
   Graphical methods
   Least squares methods
   Analysis of fit lines/curves
   Correlations and other measures of relations
   Use and misuse of statistics

Uncertainty and Probability

Informal likelihoods and vocabulary
Numerical probability and probability models
Counting as it applies to probability
Mutually exclusive events
Conditional probability & independent events
   Independent events
   Conditional probability

Bayes' theorem
Contingency tables
Probability distributions for discrete random variables
Probability distributions for continuous random variables
Expectation and the algebra of expectations
Sampling and probability
Estimation of population parameters

Hypothesis testing
Confidence intervals
Bivariate distributions
Markov processes
Monte Carlo methods and computer simulations

Elementary Analysis

Infinite Processes

Arithmetic, Geometric, and General Sequences
   Arithmetic sequences: n^{th} terms
   Geometric sequences: n^{th} terms
   General sequences: n^{th} terms
Arithmetic, Geometric, and General Series
   Arithmetic sequences: sums
   Geometric sequences: sums
   General sequences: sums

Binomial Theorem
   Binomial series
Other Sequences and Series
   Patterns to rules
   Difference equations description
Limits and Convergence of Sequences & Series
   Limits/convergence of sequences
   Limits/convergence of series
   Maclaurin and Taylor series
   Hyperbolic trigonometric series

Limits and functions
   Limit of function as $x \rightarrow$ a
   Limits at infinity
   Limits for sequence of functions
   Theorems about limits
Continuity
   Concept and definition
   Intermediate value theorem

Change

Growth and Decay
Differentiation
Concept and definition (algebraic & geometric)
Derivative of power functions
Derivative of elementary functions
Derivatives of sums, products, and quotients
Derivatives of composite functions (Chain rule)
Derivatives of implicitly defined functions
Derivatives of higher order
Relationship between derivative behavior and maxima and minima
Relationship between derivative behavior and concavity and inflection points

Relationship between differentiability and continuity
Mean value theorem
l'Hôpital's Rule
Applications of the derivative
Derivative of polar function
Derivative of vector valued functions
Newton's method
Integration
Concept and definition
Basic integration formulas
Integration by substitution
Integration by parts
Integration by trig substitution
Integration by partial fractions

Definite integrals—limit of sums
Properties of integral
Approximations of definite integral
Fundamental Theorems
Applications of definite integral
Antiderivatives
Integration of polar functions

Differential Equations
Partial Differentiation
Numerical Analysis Considerations
Multiple Integration

Validation and Justification

Logical connectives
Quantifiers (For each, there exists, ...)
Boolean algebra

Validation and Structure

Conditional statements / equivalence of statements (converse, contrapositive, inverse,....)
Inference schemes (e.g., modus ponens, modus tolens, syllogism,....)

Direct deductive proofs
Indirect proofs and proof by contradiction
Proof by mathematical induction
Consistency and independence of axiom systems

Structuring and Abstracting

Sets, set notation, and set operations
Equivalence relations, partitions, and classes
Groups
Rings

Fields
Vector spaces
Subgroups, subspaces, and their properties
Other axiomatic systems (finite geometries,....)

Isomorphism
Homomorphism

Augmented TIMSS Curriculum Frameworks

Other Content

Informatics		
Operation of computers Flow charts	Programming language Programs	Algorithms with application to the computer
History and Nature of Mathematics		
Special Applications of Mathematics		
Kinematics Netowian mechanics	Networks (graph theory) Linear programming	Critical path analysis Econometrics
Problem Solving Heuristics		
Non-Mathematical Non-Science Content (association of mathematics with content and actions in non-science area)		

Appendix B

Technical Notes

Cautions About Unwarranted Conclusions

Guarding against overgeneralization of results. Reiterating a caution provided in Chapter 1, study results should not be overgeneralized into statements that a given country's curricula or examinations include (or do not include) specific topics because: (1) only two years of examinations were analyzed; we have no specific knowledge of topics in other years of the examinations; (2) while examinations in other countries undoubtedly have tremendous influence on the topics studied in school, research on the specific linkages between examination topics and school topics has not been done. Because the structure of examinations probably is more stable than their topics over the years, cautious generalizations about length, choice, item type and performance expectations are more appropriate than about specific content.

Countries not included in the study. Resource constraints limited the study to seven countries, although many others have university entrance examinations. The included countries generally were chosen because they are strategic economic partners of the United States, and we had preliminary information that there were interesting contrasts between their examinations and the Advanced Placement examinations with respect to both their internal characteristics and their examination systems. We do not know whether examinations of some other countries would have similar or even more striking contrasts.

Percentages of Topic Coverage in England/Wales examinations. Because the England/Wales examinations are so long, 1.5 to 3 times longer than examinations from the other countries, seemingly small percentages of topic coverages in the England/Wales examinations actually represent significant amounts. Only a few percent of an England/Wales examination can represent several examination questions; in contrast, a few percent of other countries' examinations represent only one or two questions.

Information on Methods

Topic stability between years. We included two years of examinations and reported their aggregate characteristics in order to provide as much generalization about countries examinations as possible. There was some variance in examination topics between years, but overall there was more stability than variance for the more general topics. More between-year variance existed for very detailed topics. Tamir and Dossey have provided analyses for year-to-year stability at a general topic level for biology and mathematics examinations. Biology: Except for the Associated Examining Board in England and Wales, topic stability ranged from 60 to 95 percent, and averaged 80 percent. Mathematics: Pearson product-moment correlations ranged from 0.64 to .98 except for the examinations of Aix (France), Sweden, and Tokyo University, which each had little commonality among topics across the two years. Because these examinations contain a small number of questions, changes in the topics of only a few questions can result in large variances in these examinations' topic coverages.

Analytical techniques in Chapter 6, mathematics. The comparisons in Chapter 6 are supported by especially extensive mathematical analyses that the editors mostly omitted to make results accessible to as many readers as possible. For example, averages for topic coverage over two years were subjected to a median polish (Hoaglin, Mosteller, and Tukey, 1983). Topic patterns among countries were derived from a cluster analysis (SAS Institute, 1994). Further analysis and description of the clusters was done through dendograms that were not included in the chapter (Milligan, 1980).

Distinction between short and extended answers. Short answers were defined to be 1-3 sentences of text or quantitative answers requiring only one formula or equation in a single-calculation step. Extended answers were identified as four or more sentences of text or quantitative answers requiring multiple-calculation steps and/or more than one formula or equation. Although authors encountered some ambiguity between short- and extended- answer questions within their subjects, they felt the distinction was sufficiently clear to report these two item types separately in Chapters 3 through 6. Because this delineation between short and extended answers was not always easy to maintain in similar ways among the different subjects, however, Chapter 2 discusses the study's findings about these types of items in the aggregate.

Within-country reporting for England/Wales, France, and Germany. The examinations from the Associated Examining Board and the University of London usually were different enough to warrant separate reporting of their characteristics in many tables and figures. Because differences between the Aix and Paris examinations in France and the Baden-Württemberg and Barvaria examinations in Germany were less pronounced, often the aggregate data for France and Germany are reported. However, readers will find some separate reporting of French and German regions for examination topics.

Scorable events. The unit of analysis in the study was scorable events as described in Chapter 1—the smallest question in an examination that could not be broken down into more subquestions. Many examinations had questions that were numbered as a single item yet had several subquestions embedded within them. The authors analyzed each subquestion separately, i.e., these were the scorable events for coding, analysis and reporting. The data in tables or figures were compiled from the coding of scorable events.

Estimated weightings for scorable events. While many examinations provided the points allocated to whole questions or subquestions, as they were numbered by the examinations themselves, we often had to estimate the points corresponding to the embedded subquestions that we identified as scorable events. Sometimes this information could be gleaned from scoring rubrics. When scoring rubrics were unavailable, however, we had to divide the available points evenly across the scorable events, or estimate a weighting if the scorable events embedded in a question obviously required different amounts of student effort, e.g., one scorable event required a short answer while the other necessitated an extended answer.

Influence of choice on weightings. When students were afforded a choice among questions, each scorable event was weighted accordingly. For example, if students were to answer any three of eight questions, then the contribution of each scorable event when compiling the examinations' topics, etc., was multiplied by 3/8.

Terms in Text

Scorable events, items, questions. All specific data reported are referring to compilations of *scorable events*. To make the text more accessible to readers, however, authors often used more common words—*items* or *questions.*

Papers, sections. Most examinations contained discrete subparts that are described in Chapters 3 to 6. Parts that are separately timed are called *papers*, while parts with no formal timing requirements are referred to as *sections*. This particular usage was adopted in part because examinations in England and Wales have separately timed parts that are called "Papers."

Points, marks, grades, scores. Examination questions have assigned values that we refer to as points. We used the term *scoring* for referencing the process of awarding points for students' answers to individual questions. The total number of points given for a student's performance on the examination was called the *score. Grading* is the process of translating a total examination score into some reporting scale such the AP scale of 1-5. The resulting value assigned is called the examination *grade.* Obviously, these terms are often used in alternative ways, or interchangeably—so much so, that we may have inadvertently failed to be consistent with our intended use of them in the book. A final confusion is that in England and Wales, the points awarded for individual questions are called *marks* and the process of scoring the examinations is called *marking.*

References

Advanced Placement Program (AP Program)

1992 *The 1990 Advanced Placement Examination in Biology and Its Grading.* New York: The College Board.

1993 *A Guide to the Advanced Placement Program.* New York: The College Board.

1994 *1994 AP Year Book.* New York: The College Board.

American Association for the Advancement of Science (AAAS)

1989 *Project 2061: Science for All Americans.* Washington, DC: author.

1993 *Benchmarks for Science Literacy.* Washington, DC: author.

American Chemical Society (ACS)

1993 *ChemCom: Chemistry in the Community.* 2nd ed. Dubuque, IA: Kendall/Hunt.

August, R.

1992 "Yobiko: Prep School for College Entrance in Japan." In R. Leestma and H. Walberg, eds., *Japanese Educational Productivity.* Ann Arbor, MI: University of Michigan.

Black, P.

1992 *Physics Examinations for University Entrance.* New York: UNESCO.

1994 "Performance Assessment and Accountability: The Experience in England and Wales." *Educational Evaluation and Policy Analysis.* 16(2):191-203.

Bloom, B.

1956 "Taxonomy of Educational Objectives: The Classification of Educational Goals." In *Handbook I: Cognitive domain.* New York: Longman.

Bodin, A.

1994 "Note sur l'Examen du Baccalauréat (FRANCE)." Unpublished paper. Université Franche Comte.

Centre for Educational Research and Innovation

1993 *Education at a Glance: OECD.* Paris: Organisation for Economic Co-operation and Development.

1995 *Education at a Glance: OECD.* Paris: Organisation for Economic Co-operation and Development.

College Board, The

1994a *Advanced Placement Course Description: Mathematics: Calculus AB, Calculus, BC.* New York: The College Board.

271

1994b *Advanced Placement Course Description: Physics: Physics B, Physics C.* New York: The College Board.

Colomb, J.
1995 "Pupil Assessment and Examination in Secondary Education." In E. Kangasniemi and S. Takala, eds., *Pupil Assessment and the Roles of Final Examinations in Secondary Education.* Lisse, the Netherlands: Swets and Zeitlinger B.V.

Darling-Hammond, L.
1991 "The Implications of Testing Policy for Quality and Equality." *Phi Delta Kappan,* 73(3):220-225.

Dobson, K.
1992 "The A-Level Examination in the U.K." In P. Black, ed., *Physics Examinations for University Entrance.* New York: UNESCO.

Doran, R., Lawrenz, F., and Helgeson, S.
1994 "Research on Assessment in Science." In D. Gabel, ed., *Handbook of Research on Science Testing and Learning.* New York: Macmillan.

Dossey, J.
1994 *"Content Analysis of Mathematics Examinations."* Paper presented at annual meeting of the American Educational Research Association, New Orleans, April.

Eckstein, M., and Noah, H.
1993 *Secondary School Examinations: International Perspectives on Policies and Practice.* New Haven: Yale University Press.

Eckstein, M., and Noah, H., eds.
1992 *Examinations: Comparative and International Studies.* Oxford, England: Pergamon Press.

Embassy of France
1991 *Organization of the French Educational System Leading to the French Baccalauréat.* Washington, DC: Embassy of France.

Evans, K.
1994 "Change and Prospects in Education for Young Adults." *Comparative Education,* 30(1):39-47.

Fägerlind, I.
1992 "Beyond Examinations: The Swedish Experience and Lessons from other Nations." In M. Eckstein and H. Noah, eds., *Examinations: Comparative and International Studies.* Oxford, England: Pergamon Press.

Feuer, M., and Fulton, K.
1994 "Educational Testing Abroad and Lessons for the United States." *Educational Measurement: Issues and Practice,* 13(2):31-39.

Frost, P.
  1992   "Tinkering With Hell: Efforts to Reform Current Japanese University Entrance
         Examinations." In M. Eckstein and H. Noah, eds., *Examinations: Comparative and
         International Studies.* Oxford, England: Pergamon Press.
Führ, C.
  1989   *Schools and Institutions of Higher Education in the Federal Republic of Germany.*
         Bonn: Inter Nationes.

Gandal, M., Britton, E., Hawkins, S., Hokanson, C., Wattenberg, R., and Raizen, S.
  1994   *What College-Bound Students Abroad Are Expected To Know About Biology.*
         Washington, DC: American Federation of Teachers and National Center for
         Improving Science Education.

Gisselberg, K. and Johansson, G.
  1992   "Sweden: School Assessments and Central Tests." In P. Black, ed., *Physics
         Examinations for University Entrance.* New York: UNESCO.

Halls, W.
  1994   "16-19: Some Reflections on Europe and the Reforms." *Comparative Education,*
         30(1):25-29.

Herr, N.
  1992   "A Comparative Analysis of the Perceived Influence of Advanced Placement and
         Honors Programs upon Science Instruction." *Journal of Research in Science
         Teaching,* 29(5):521-532.

Heyneman, S. and Fägerlind, I., eds.
  1988   University Examinations and Standardized Testing. Washington, DC: The
         International Bank for Reconstruction and Development.

Hoaglin, D., Mosteller, F. and Tukey, J., Eds.,
  1983   *Understanding Robust and Exploratory Data Analysis.* New York: Wiley.

Kelly, P.
  1994   "Biological Education in the United Kingdom." In P. McWethy, ed., *Basic
         Biological Concepts: What Should the World's Children Know?* Reston, VA:
         National Association of Biology Teachers.

Kelly, P. and Lister, R.
  1969   "Assessing Practical Ability in Nuffield A-Level Biology." In J. Eggleston and J.
         Kerr, eds., *Studies of Assessment.* London: English University Press.

Kesner, M., Hofstein, A., and Ben-Zvi, R.
  1995   *"How to Make Chemistry More Relevant to the Students."* Paper presented at the
         annual meeting of the National Association for Research in Science Teaching, San
         Francisco, CA, April.

Krusemark, D. and Forsaith, A.
  1995   *What Secondary Students Abroad Are Expected to Know: Gateway Exams Taken by
         Average-achieving Students in France, Germany, and Scotland.* Washington, DC:
         American Federation of Teachers.

Langlois, F.
  1992 "The French System of Entrance Examinations." In Paul Black, ed., *Physics Examinations for University Entrance*. New York: UNESCO.

Lazarowitz, R. and Tamir, P.
  1994 "Research on Using Laboratory Instruction." In D. Gabel, ed., *Handbook of Research on Science Testing and Learning*. New York: Macmillan.

Leestma, R., et al.
  1987 *Japanese Education Today*. Washington, DC: United States Department of Education.

Linn, M.
  1987 "Establishing a Research Base for Science Education: Challenges, Trends and Recommendations." *Journal of Research in Science Teaching*, 24(3):191-216.

Madaus, G., and Kellaghan, T.
  1991 *"Student Examination Systems in the European Community: Lessons for the United States."* Contractor report submitted to the Office of Technology Assessment. Washington, DC.

Masters, B.
  1994 "England's New Class System." *The Washington Post Education Review*, Nov. 6.

Milligan, G.
  1980 "An Examination of the Effect of Six Types of Error Perturbation of Fifteen Clustering Algorithms." *Psychometrika*, 45:325-342.

Ministry of Education, Science, and Culture, Government of Japan (Monbusho)
  1994 Monograph by the Ministry of Education, Science and Culture. Tokyo: author.

Nakayama, K.
  1994 "Biology Teaching and Its Components in Japan." In Patricia McWethy, ed., *Basic Biological Concepts: What Should the World's Children Know?* Reston, VA: National Association of Biology Teachers.

National Council of Teachers of Mathematics
  1989 *Curriculum and Evaluation Standards for School Mathematics*. Reston, VA: Author.

National Center for Educational Statistics (NCES)
  1993 *Digest of Education Statistics: 1993*. Washington, DC: United States Government Printing Office.

National Center for Improving Science Education
  1991 *The High Stakes of High School Science*. Andover, MA: The NETWORK, Inc.

National Endowment for the Humanities
  1991 *National Tests: What Other Countries Expect Their Students to Know*. Washington, DC.: National Endowment for the Humanities.

National Research Council (NRC)
  1989 *High School Biology Today and Tomorrow.* Washington, DC: National Academy Press.
  1994 *National Science Education Standards.* Draft. Washington, DC: National Academy Press.

National Science Teachers Association (NSTA)
  1992 *The Content Core: A Guide for Curriculum Designers.* Washington, DC: National Science Teachers Association.

Nolan, K.
  1994 "Mathematics in France, 1994." New Standards Project. Pittsburgh, PA: Learning Research and Development Center, University of Pittsburgh.

Resnick, L., Nolan, K., and Resnick, D.
  1994 *Benchmarking Education Standards.* Washington, DC: National Research Council.

Robitaille, D., Schmidt, W., Raizen, S., McKnight, C., Britton, E., and Nicol, C.
  1993 *Curriculum Frameworks for Mathematics and Science.* Vancouver, Canada: Pacific Educational Press.

Ryu, T.
  1992 "The University Tests in Japan." In Paul Black, ed., *Physics Examinations for University Entrance.* New York: UNESCO.

SAS Institute
  1994 *JMP-Statistics and Graphics Guide.* Cary, NC: SAS Institute.

Shimahara, N.
  1992 "Overview of Japanese Education: Policy, Structure, and Current Issues." In R. Leestma and H. Walberg, eds., *Japanese Educational Productivity.* Ann Arbor, MI: University of Michigan.

Shimizu, K.
  1994 Private communication. Tokyo: National Institute for Educational Research.

Showalter, W.
  1986 *Conditions for Good Science Teaching.* Washington, DC: National Science Teachers Association.

Shulman, L. and Tamir, P.
  1973 "Research on Teaching in the Natural Sciences." In R. Travers, ed., *Second Handbook of Research on Teaching.* Chicago: Rand McNally.

Stevenson, H., et al.
  1994 *International Comparisons of Entrance and Exit Examinations: Japan, United Kingdom, France, and Germany.* Draft. Washington, DC: United States Department of Education.

Tamir, P.
  1972 "The Role of the Oral Examinations in Biology." *School Science Review*, 54:162-175.

  1974 "An Inquiry Oriented Laboratory Examination." *Journal of Educational Measurement*, 11(1):25-33.

  1975 "Nurturing the Practical Mode." *The School Review*, 83, May: 499-506.

  1985 "The Israeli 'Bagrut' Examinations in Biology Revisited." *Journal of Research in Science Teaching*, 22(1):31-40.
  1993 *"Student Assessment: Present and Future Trends."* Invited Plenary Lecture, Fifth European Meeting of Biology Teaching, Barcelona, May.

  1995 Personal communication.

Thissen, D., Wainer, H., and Wang, X.
  1994 "Are Tests Comprising Both Multiple-Choice and Free-Response Items Necessarily Less Unidimensional Than Multiple-Choice Tests? An Analysis of Two Tests." *Journal of Educational Measurement*, 31(2):113-123.

Wu, Ling-Erl E.
  1993 *Japanese University Entrance Examination Problems in Mathematics*. Washington, DC: Mathematical Association of America.

Index of Sample Examination Questions